Handbook
to the
Rogue River Canyon

By:
James M. Quinn,
James W. Quinn
James G. King

Frank Amato Publications
Portland, Oregon

This book is dedicated to all people who acknowledge the gift of life and live it to its fullest...

Other books available by this book's authors are:

HANDBOOK TO THE DESCHUTES RIVER CANYON

HANDBOOK TO THE ILLINOIS RIVER CANYON

HANDBOOK TO THE MIDDLE FORK OF THE SALMON RIVER

Additional copies of the Guide Handbooks may be
ordered from:

Frank Amato Publications, Inc.
P. O. Box 82112
Portland, Oregon 97282
(503) 653-8108

FIRST EDITION — 1978
SECOND EDITION — 1979, 1980, 1983, 1985, 1987
THIRD EDITION — 1995

Library of Congress Cataloging in Publication Data
Catalog No. 77-91410

ISBN: 1-878175-50-5
Printed in HONG KONG

CONTENTS

RIVER TRIP PERMITS are required before floating the Rogue between Grave Creek and Watson Creek (regulated area) from the Friday preceding Memorial Day through Labor Day (the regulated season). Noncommercial reservations (private) will permit a maximum of sixty-two (62) persons per day to enter the regulated area.

TRIP APPLICATION forms and information may be obtained from:

Galice Ranger District
P. O. Box 1131
Grants Pass, Oregon 97526

Phone (503) 479-3735

WATER FLOW and TEMPERATURE information may be obtained from:

Grants Pass Water District
Phone (503) 476-8801, ext. 25 weekdays
(503) 476-8806 holidays & weekends

BACKFERRY   -Row backward to pull away from an obstacle.

PULL OARS IN   -Fold the oars back along the side of the boat or raft to get through a narrow area.

STANDING WAVE   -A fixed wave where fast water passes over a submerged rock.

Shuttle service may be arranged at the Galice Resort.
Phone (503) 476-3818

ACKNOWLEDGEMENTS

Although considerable time has gone into the creation of this book, it has been a labor of love. Many hours, days and years of floating have been involved, each moment of which has been a delight.

Many friends and associates have contributed knowingly or unknowingly to this work. For instance, our frequent river companions, Dave Boals, Jack Ingram and Ralph Hibbs, by their constant harrassment and badgering, have stimulated us to learn where the rocks are so that we may miss most of them. The elder author wishes to express appreciation to Tom Glatte, who risked life and limb to accompany him on his first trip down the river but who would never do it again.

Special appreciation and gratitude are extended to Glen Wooldridge, the dean of river boatmen. Glen has done it all and he did it all first. He has allowed us to examine his personal files, let us make copies of his pictures and photos and given freely of a vast store of information covering 60 years of running the Rogue River.

We want to express our thanks to James G. King, who has been our special photographer and given us invaluable technical aid in preparing this material for printing.

Several of our friends, who are professional guides, have examined the river log and offered helpful comments and advice. These include Bob Pruitt, Jim Wallis, Willie Illingworth, Paul Brown, Jerry Pringle and Mike Saul. We sincerely thank them all.

INTRODUCTION

If you love whitewater, if you enjoy a challenge, if you feel close to nature and if you love the land, you are sure to delight in your adventure down the Rogue River Canyon. The lower Rogue offers one of the wildest boat rides in the Northwest. After you have completed your journey, you are not likely to disagree.

The whole thing started about a million years ago, when the western part of the state was uplifted to form the Pacific Coast Range. It was during this activity that the lower Rogue River was transformed from a quiet stream to a frothing, boiling torrent of whitewater. As the land slowly rose, the river eroded its channel deeper and deeper into the rock until today the canyon averages a depth of over 3,000 feet. As you start your journey into the past at Grave Creek, the elevation is approximately 690 feet above sea-level. After you complete your voyage at Foster Bar, the elevation has dropped to 155 feet over the distance of 34 miles. The average drop per mile is about 15 feet, but the gradient ranges from over 30 feet per mile in Mule Creek Canyon, to 3 feet per mile in the Clay Hill Stillwater section.

A boat which starts at Grave Creek plunges into the whitewater in the steep-sided lower canyon. Confined between narrow rock walls, the river savagely flings itself against every ledge and boulder that confronts it. In a distance of less than twenty miles, there are over 40 named rapids, and in one place (Rainie Falls) the entire river plunges over a twelve foot perpendicular ledge. The wildest part of the journey, and probably the most exciting continuous flow of rapids, occurs between Rainie Falls and Black Bar. This section of the river will challenge the most able boatman, as one rapid transforms into another. At Mule Creek Canyon the river dissects the heart of the Coast Range. For over a mile the whitewater swirls and boils through a deep, narrow trough. The Rogue relaxes after the churning cauldron of Mule Creek Canyon, and you may feel the river has finally been tamed. However, within twenty minutes of downstream drifting you will enter an unprecedented experience in the rock-garden known as Blossom Bar. The water swirls and churns through giant boulders which will destroy any type of craft. Finally, below Blossom Bar, the river begins to slow down and widen. The rapids are fewer and farther apart. This area invites you to relax as you slowly float downriver.

Thousands of people each year enjoy the Rogue River Canyon between Grave Creek and Foster Bar. Many come down the river in kayaks, rafts, drift boats, inflatables and on foot. Most of this activity occurs during the summer months, from Memorial Day to Labor Day. After the first of September the steelhead fishermen are often the only people on the river.

In 1978 the Bureau of Land Management restricted the number of private parties journeying down the river. Since 1974, the number of commercial parties using the river has also been restricted. As the pressure of human traffic increases along the river, campsites overload, firewood becomes scarcer and it becomes necessary to carry portable stoves or cooking devices. All travelers are requested to leave no trash or garbage at any camp-site. Carry all cans and bottles out with you, because if they are buried the flooding of the river will still expose them. Use no more wood than you need. If you like to barbeque your food, bring your own charcoal. Whenever possible use dehydrated foods, avoid canned-goods as much as you can to minimize garbage residue.

The river flow is now regulated to a considerable degree by Lost Creek Dam. Summer flows may vary from 1200 to 2500 cubic feet of water per second. The difficulty of the river varies with its water volume. During flood stages the volume may drastically increase, causing this river to be almost impossible for navigation. In 1964 flood, the Rogue reached 500,000 cubic feet per second at Agness.

BOATING SAFETY

Before starting down the river you must know your limitations and abilities. Your age, physical condition and experience should all enter into your decisions. Never overload your boat so that it cannot be turned quickly and easily maneuvered. Load the boat with the heavy objects near the bottom to keep the center of gravity at its lowest point (to reduce the likelihood of swamping). After properly loading the boat, securely tie down all objects in the boat to prevent weight shift. Make sure the boat, including gear and passengers, is balanced perfectly before entering any section of the river which necessitates maneuvering. Assign specific positions to passengers to prevent loss of balance. Be sure you and your passengers know what to expect from each other during emergencies. Passengers must have their lifejackets on and securely fastened. Once started down the river you cannot backtrack easily, and help may be many miles away. Don't take chances. Any rapids which cannot be clearly seen ahead should be scouted before entering. Pull ashore upstream and evaluate the rapids carefully. Predetermine your route through the rapids before returning to your boat. Never be overconfident in any rapids no matter how small it may appear. The least obvious factor can cause extreme difficulties at any time. It is important to realize that hazards vary with the river level — during high water, standing waves may appear, creating impassable suckholes and back-eddies. Sometimes a new course will have to be found. For example, Tyee Rapids may safely be negotiated on the right side of the river during low water, but during high water the only safe channel is on the far left bank. When passing through whitewater, the passengers should be assigned specific parts of the boat to grasp to prevent uncontrolled weight shifts. Never get broadside to the current in fast water. If you see that you are going to hit an obstruction, NEVER hit it broadside, ALWAYS hit it head on. Always counterbalance the boat when striking an object — this can be accomplished by shifting the weight to the 'high' side (the opposite side to which the boat is leaning). — this should be practiced before your journey. Don't wait until an emergency to find out that your passenger doesn't know how to counterbalance! Steer clear of overhanging branches and all submerged objects. The current's force is much stronger than any human, and must be respected at all times — it can cause serious injuries and accidents. Don't overexert your physical capacities — don't try to float too far in a single day. Fatigue sets in slowly and may weaken you when you need maximum strength and alertness.

Your boat should be outfitted with the following essential pieces of equipment: one 100 ft. lining rope for lining your boat around the falls, severe rapids, and nighttime securing when fastenings are some distance from the river; adequate rope for tie-downs to secure your equipment inside your boat (we recommend four 25 ft. lengths of ¼ inch rope — these can be tied together to make a second 100 ft. rope for emergency situations); two extra oars or paddles (oars will break — sometimes in the middle of rapids) position one on either side of your boat, readily available in case of emergency; two extra oarlocks (oarlocks may break or be lost by accident — they don't float) secured to the boat by wire; a boat repair kit suited to repair your type of craft; a plastic bailing bucket (conforms to the contour of the bottom of your boat) to remove water which will come aboard during passage through rapids; (water weighs about 8 lbs per gallon, so you must keep it where it belongs, in the river, not the boat).

Check your boat before your journey. Be sure it is in good physical condition. Make sure wooden boats do not have dry rot; fiberglass boats should not be cracked or leaking; rubber boats should not have leaks or cracks and the rubber should be in good shape. Be certain the blocks that hold the oarlocks in place are not cracked or split and are securely attached to the frame of the boat — all the stress of the rapids will be transmitted through these sensitive areas and they are the primary source of problems when accidents occur on the river.

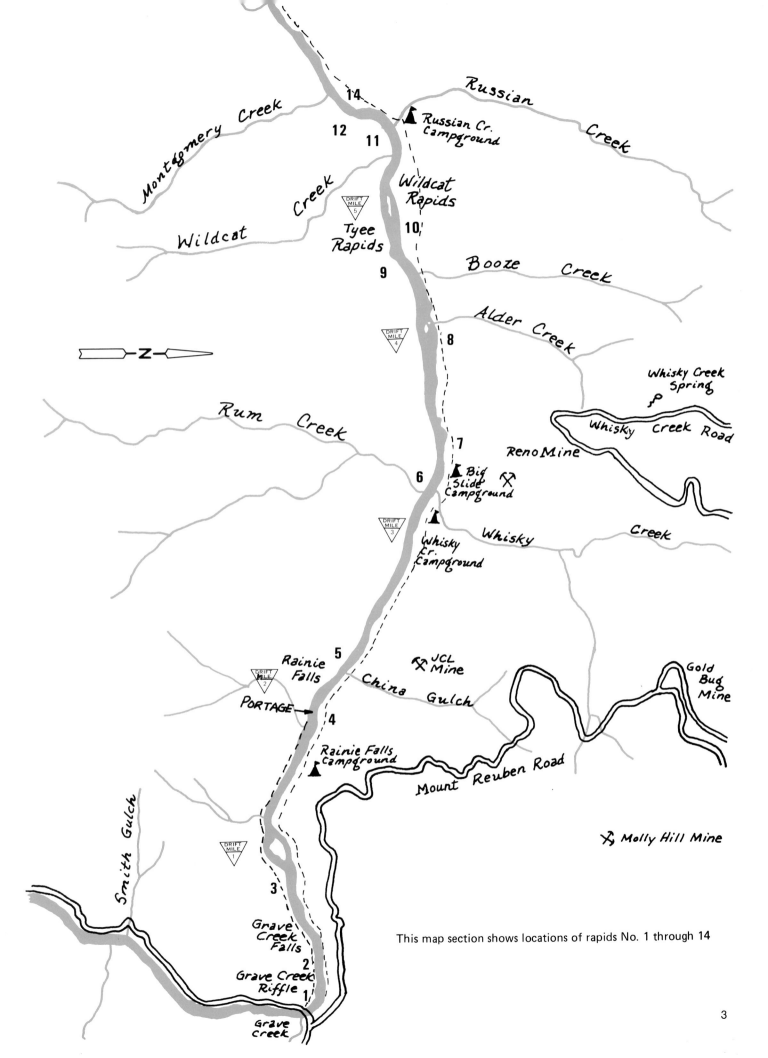

Montgomery Creek

Russian Creek

14

12

11

🏕 Russian Cr. Campground

Wildcat Creek

△ DRIFT MILE 5

Tyee Rapids

Wildcat Rapids

10

Wildcat

9

Booze Creek

Alder Creek

8

△ DRIFT MILE 4

Whisky Creek Spring

Whisky Creek Road

Rum Creek

7

Reno Mine

6

🏕 Big Slide Campground ⚒

△ DRIFT MILE 3

Whisky Creek

🏕 Whisky Cr. Campground

⚒ JCL Mine

5

Rainie Falls

China Gulch

Gold Bug Mine

△ DRIFT MILE 2

PORTAGE ➔

4

🏕 Rainie Falls Campground

Mount Reuben Road

⚒ Molly Hill Mine

Smith Gulch

△ DRIFT MILE 1

3

Grave Creek Falls

This map section shows locations of rapids No. 1 through 14

2

Grave Creek Riffle

1

Grave Creek

3

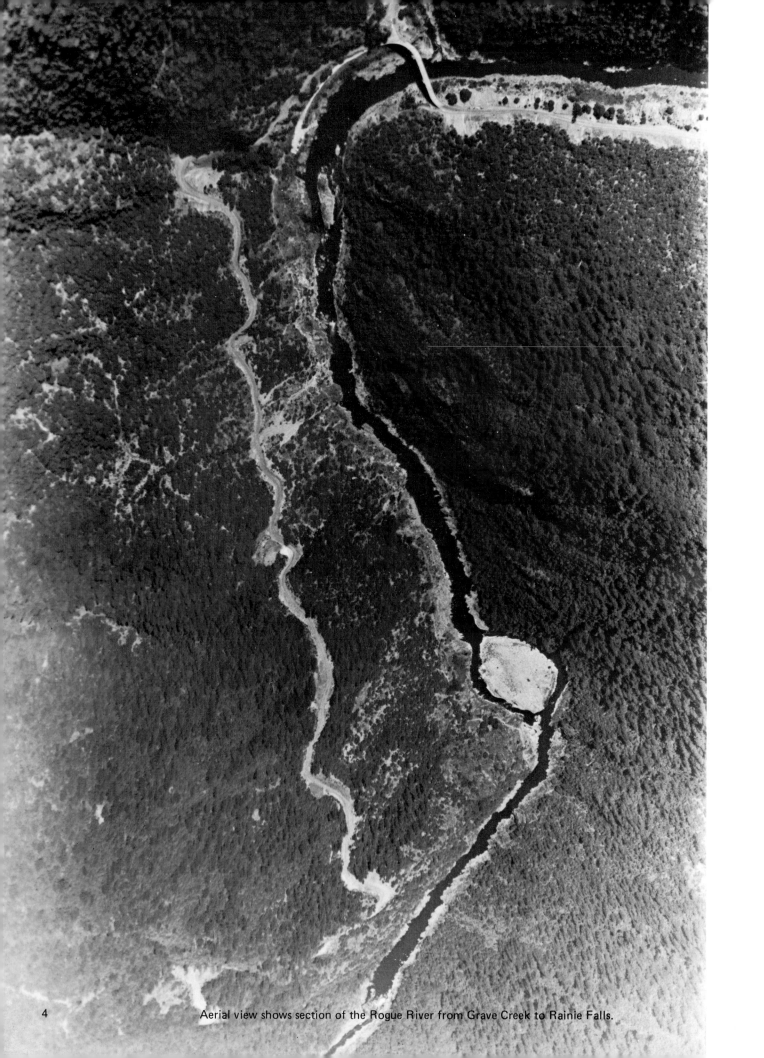

Aerial view shows section of the Rogue River from Grave Creek to Rainie Falls.

Rapid No.	Drift Mileage	Time	Class	Name	
---	0.0	---	---	GRAVE CREEK LANDING	After traveling some 7 miles downstream from Galice we cross the Rogue River bridge at Grave Creek. Immediately after crossing the bridge, turn left and down to the river where the boat ramp is located. This boat ramp identifies the end of the road and the beginning of the Rogue River Hiker's Trail. The U.S. Bureau of Land Management (B.L.M.) maintains a check-station here for the registration of boaters and hikers. Outdoor restroom facilities are available. After registering your boating party, you are ready to begin your journey down the most popular whitewater river in Oregon. Grave Creek was named because Martha Leland Crowley, daughter of a pioneer couple, was buried under a prominent oak tree near the stream in 1846. The community of Leland was named in her honor, and in 1854, an unsuccessful attempt was made to change the name of Grave Creek to Leland Creek. There has never been a road into the heart of the canyon. In pioneer days the Indian trails were the only means of access and egress for the hardy prospectors and rivermen. During the 1850's two pioneers, named Bates and Twogood, operated a ranch on Grave Creek and packed supplies in to the miners of this area. In the winter of 1852-53, when snow blocked all passages and trails, a pound of flour was worth one dollar, an ounce of tobacco cost one dollar and salt was traded even weight for gold!

Grave Creek Rapid — note that the main channel is on the south side of the island.

Rapid No.	Drift Mileage	Time	Class	Name
1	0.1	1 min	III	GRAVE CREEK RAPIDS

Identified as the first whitewater of your journey, only a few hundred feet downstream from the launching site. The main channel sweeps left around a large rock/gravel bar. At high water levels both sides of the gravel bar are navigable. As the water level drops in summer the boater must shoot the left channel.

HOW TO: Approach the rapids to the RIGHT OF CENTER. Continue down through the whitewater by maintaining position slightly LEFT OF CENTER of the river. At the LOWER END of the rapids (approximately 150 feet downstream from its beginning) the river takes a slight bend to the right. At this bend, the river forms a BACK-EDDY ON THE LEFT side which can trap you against the rock ledge located there. Also at this bend, in the middle of the stream, there is a large boulder which must be avoided. PASS BETWEEN THE LEFT BANK AND THE MIDSTREAM BOULDER, AND KEEP OUT OF THE BACK-EDDY. When Zane Grey made his first trip down the Rogue, his boat was trapped in the back-eddy mentioned here, and his boat was punctured. The hazard is still present.

Grave Creek Rapid — this section can be coordinated with the 'HOW TO'

Rapid No.	Drift Mileage	Time	Class	Name	
2	0.2	1 min	III	**LOWER GRAVE CREEK FALLS**	Identified only a few hundred feet farther downstream from Grave Creek Rapids.

HOW TO: Approach the falls by positioning yourself approximately 1/3 FROM THE LEFT BANK. In the channel a submerged boulder can be seen in the approximate midstream. PASS BETWEEN THE LEFT BANK AND THE MIDSTREAM BOULDER. Maintain this position as you pass through the falls, which has a vertical drop of 3-5 feet (vertical drop depends upon the water-level of the river).

| | 0.6 | | | SANDERSON'S CABIN | A concrete foundation of an old cabin is situated on the north bank. |

Potable water is available from a small spring near the foundation. Glen Wooldridge advises there were three Sanderson brothers who arrived here from Ohio in 1903. The presently-visible foundation was built in 1940. The Sanderson brothers discovered a large mine up Whiskey Creek. The only access to this area was by trail and river. In 1971 the B.L.M. dismantled the cabin, leaving only these concrete remnants as a reminder of many years of development by the Sandersons.

Lower Grave Creek Falls.

Rapid No.	Drift Mileage Time	Class	Name	
3	0.9	8 min	II	**SANDER-SON'S RIFFLE**	Identified by an island (Sanderson's Island) which appears at the end of the stillwater drift. The main river channel passes to the left side of the Island.

HOW TO: Approach this main channel in EXACT MIDSTREAM and ride through on the crests of standing waves. At the LOWER END of the two-minute float past Sanderson's Island, you will find another small rapids. Its only obstruction is a rocky outcropping (in low water) located approximately 1/3 into the stream from the right bank. Therefore, position yourself approximately 1/3 FROM THE LEFT BANK as you ride through these rapids at their greatest depth. |
| --- | 1.1 | --- | --- | **SANDER-SON'S BRIDGE** | The concrete piers which are visible on both sides of the river are all that remain of an old trail-bridge which was used for mule trains and foot traffic. It was built in 1907 and destroyed by a severe flood in 1927. The bridge was built by the U.S. Government and, according to Glen Wool-dridge, was seldom used by local miners. Up on the south side of the river, behind this landmark, you can see the remains of an old miner's cabin, high on the mountainside. At a lower level you can identify the short trail extending from Grave Creek to Rainie Falls. |
| --- | 1.7 | --- | --- | **RAINIE FALLS CAMPSITE** | On the left bank, just above Rainie Falls, a good campsite may be found. Pull your boats onto the sandy beach. A nice place to eat a snack before lining the fish ladder. Fresh water can be found in the small spring at the downriver end of the site. Another campsite is located on the right side of the river, at the head of the fish ladder. |

Lining the Drift Boat down the fish-ladder at Rainie Falls. Be cautious — rocks are slippery here.

A group of small inflatables enjoys a wild ride down the fish ladder at Rainie falls

9

Rapid No.	Drift Mileage	Time	Class	Name
4	1.8	12 min	VI	RAINIE FALLS

This Falls was named after old man Rainie, who lived in a small cabin below the Falls, and made his living by gaffing salmon at the Falls. He took his catch out by horseback and sold it at Glendale. At this area of the Rogue River Canyon, the river comes into a highly resistant type of rock which has not given way to erosion by the river as readily. Here the water is forced to pass through a narrow chute which causes an approximate 12 foot vertical drop, strong hydraulics and turbulence. These are unsafe conditions for almost every type of river craft. See note below:

HOW TO: LARGE INFLATABLES ONLY: Use the chute between the main river flow and the fish ladder. ALL OTHER RIVER CRAFT: Line your vessel down the fish ladder on the right side of the Falls. Do not attempt to take the Falls itself. While passing through the fish ladder, be careful because of currents and slippery rock surfaces.

Rainie Falls - Note also middle channel and fish ladder on right side.

Large rafts may pass through the middle channel at Rainie Falls.

Old man Rainie (left) and unidentified friend (right). Rainie earned a living for many years gaffing salmon at Rainie Falls. During the early days, the Falls were alive with salmon at almost all times of the year.

Aerial view of the Rogue River Canyon below
Rainie Falls. This section of the river is known
as China Gulch.

Rapid No.	Mileage	Drift Time	Class	Name	
5	2.1	4 min	II	**CHINA GULCH RAPIDS**	China Creek enters on the right side. Rock structures on both banks narrow the river to form this minor rapids. This gulch was named after the Chinese miners who worked in this area in the late 1800's.
					HOW TO: Negotiate by holding in the MAIN SLOT (SLIGHTLY RIGHT OF CENTER), with small maneuvers to avoid any rocky outcroppings.
---	---	---	---	CALM WATERS	Downstream approximately 1/4 mile the calm river allows sightseers to enjoy plant and wildlife on both banks. An old-timer's cabin can be sighted on the upper levels of the left bank. Stillwater from this point until the next rapids may entice you into rowing or paddling. During this stillwater stretch, you will pass Whiskey Creek campsite, Rum Creek on the left and Whiskey Creek on the right. Near the end of the stillwater you will see dramatic exposed earthslides on the mountainsides of the north bank. These slides caused Big Slide Riffle (ahead).
---	3.1	---	---	WHISKEY CREEK CAMPSITE	This is a large campsite, suitable for the accomodation of large parties. Besides the riverside camp, there is a large campsite up the mountainside, suitable for hikers. Boaters and hikers often meet at this facility.
***	3.2	---	---	*WHISKEY CREEK CHECK-POINT*	*This is your first CHECK-POINT. The purpose of these CHECK-POINT inserts is to help you locate your position precisely as you travel down the river. You can readily determine how far you have come, and how far you have to go to reach any specific point along your journey.*
					Two mines (the JCL and the Benton) are located up Whiskey Creek a short distance. Whiskey Creek enters the Rogue on the right side (North side) of the river; Rum Creek enters the Rogue directly across the river, on its left side (South side). The water in Rum Creek is from a better watershed, resulting in its water being approximately 10 degrees cooler and fresher. There is a small sand bench above the river level on the Rum Creek side, which offers camping for small parties. Salmon can be often found schooling at the mouth of Rum Creek, because of its cooler water. If you walk a short distance up Whiskey Creek, you will find the remains of a pioneer cabin, which was built in 1880. They have been well-preserved and are an interesting feature of this area. Also, the remains of old mining equipment and ditching for water-control of hydraulic-mining are obvious at the old pioneer cabin site. These are located some 200 yards upstream on the Northwest side of the Creek.

Aerial view shows Whiskey Creek Campsite.

An old cabin located a short distance up Whiskey Creek, built by a gold-miner in the 1880's.

Chinese labor was used to dig ditches necessary to transport water to where mining was taking place. This ditching is located a short distance up Whiskey Creek.

Aerial view extending from Whiskey Creek to Tyee Bar (foreground).

Rapid No.	Mileage	Drift Time	Class	Name	
6	3.3	30 sec	II	**WHISKEY CREEK RIFFLE**	Located approximately 80 yards past the Whiskey Creek Checkpoint. **HOW TO: This minor rapids can be easily negotiated by staying IN THE CENTER of the river and remaining alert.**
7	3.5	1 min	I	**BIG SLIDE RIFFLE**	In the late 1880's, a slide occurred at this site. The entire river was blocked for some time, causing the river to back-up as far as Hell's Gate (15 miles upstream). Now the river has eroded its way through the slide, and by closely observing the banks on both sides as you pass downstream you can see the remains of the original slide. **HOW TO: Position yourself 1/3 OFF LEFT BANK. Ride through the center of the 'V'-slot.**
---	3.5	---	---	BIG SLIDE CAMPSITE	This large campsite is built on the surface of the toe of the slide. The campsite is located on the right bank, and provides area for large parties. It is necessary to haul gear up the bluff a short distance from the boats to the camp. This is primarily a trail-campsite and water should be carried from Whiskey Creek. Also, a dry campsite is located directly across the river, on the south side.
---	3.6	---	---	DOE CREEK CAMPSITE	This large, sandy beach is located on the left bank. It has water, toilets and can accomodate large groups. Its location makes it unsuitable for hikers along the trail, therefore it is used almost exclusively by river-journeyers.
8	4.0	5 min	I	**HARRY MONDALE RIFFLE**	This riffle was named after an athletic coach who has probably caught as many steelhead at this site as any other man. During President Hoover's trip down the Rogue, he caught his first steelhead on a fly at this riffle. Identify by bare rock serpentine outcroppings from both left and right banks (with several partially-submerged rocks in midstream). Looking ahead, over the riffle, you will see a rocky island some 50 yards downstream on the right side of the river. **HOW TO: This riffle involves some MINOR NAVIGATION through the partially-exposed midstream rocks. Stay in GENERAL CENTER of the river and maneuver as necessary to avoid obstructions.**
9	4.4	5 min	IV	**TYEE RAPIDS**	*****SCOUTING MANDATORY!** PULL ASHORE UPSTREAM to inspect from right bank. This rapids can be identified by the low serpentine rock outcroppings from the right bank which extend almost halfway across the river. Also, a house-sized boulder can be seen which is positioned about 100 yards downstream from the entrance to the rapids. The major rapids is formed between this large rock and the right river bank. The word 'Tyee' is the Chinook Indian word for 'chief'. Tyee Bar is the site of a famous gold-mine. It is estimated that 300 Chinese workers took over $5,000,000 worth of gold dust from this bar. Several hundred Chinese lived in the Galice area. They often would gather into parties of 20 or 30 and walk, single-file, to the stores for supplies, carrying their baskets on poles carried across their shoulders. Many guides consider this to be the most difficult rapids on the river. In 1976 a man lost his life at this site, even though he was being guided down the river. Zane Grey mentions the grave-site of a miner about 25 feet above this bar. Tyee Bar was very popular in the 1880's, when a boat-crossing and a store were operated at this site.

Tyee Rapids — this picture may be coordinated wi the the 'HOW TO' below.

USING COORDINATED RAPIDS-PLAN, IDENTIFY THE FOLLOWING:
A=pull off site; B=house rock;C=turbulence above rapids; D=submerged
boulder; E=back eddy; F=pour-off; G=main channel below rapids.

HOW TO: Just barely miss the first turbulence (C) by passing it within 5 feet on its **RIGHT SIDE**. This automatically positions you between (C) and submerged boulder (D). As soon as you have passed (D), immediately work toward the smoother water (E), within 10 feet of the right bank. Enter the extreme right pour-off (F) which is marked by a smooth 'V' slick. Ride down center of the 'V'-slick. Be careful. A huge suckhole is located on your left as you pass through the 'V'-slot. If you ride too far to right or left you will encounter difficulties. Immediately move towards midstream once out of 'V'-slot, to avoid current forcing you into rock ledge ahead. Once past the dangerous rock ledge at the lower end of the rapids, maintain position in the center of the river for a swift journey downstream several hundred feet farther. (Note: at low water levels, there are two midstream boulders which may protrude. These are located approximately 50 feet and 100 feet downstream from the 'house rock'(B). **HIGH WATER OPTION:** At levels above 2.0, you may also take the extreme left chute, staying within a few feet of the left bank at all times.

Rapid No.	Mileage	Drift Time	Class	Name
---	4.6	---	---	TYEE EMER-GENCY CAMPSITE

Parties can make an emergency camp on either side of the river immediately below Tyee Rapids. There is no water at either of these sites.

Down the slot at Tyee.

Rapid No.	Mileage	Drift Time	Class	Name
10	4.8	5 min	III	WILDCAT RAPIDS

Wildcat Rapids are formed because the Rogue splits as it passes around a gravel bar and then curves to the right. A boulder-strewn stretch of water lies at the end of the Rapids.

***SCOUT ON RIGHT BANK!

HOW TO: Use the RIGHT CHANNEL of the river. Position yourself in the EXACT CENTER OF THE RIGHT CHANNEL. Maintain this position until you reach the end of the island (approximately 250 yards), at which point the left channel rejoins the right channel and the velocity proportionately increases. The channel on the LEFT SIDE of the island is also navigable at river levels above 1.2.

Wildcat Rapids — this picture may be coordinated with the 'HOW TO' below.

USING COORDINATED RAPIDS-PLAN, IDENTIFY THE FOLLOWING: A=rocky outcroppings; B=alligator, dinosaur's backbone, or Heyerman's Reef; C=large rock outcropping; D='V'-slot; E=suckhole; F=submerged boulder.

Now that you have reached the end of the island, it gets trickier. Within 50 feet you must avoid 2 rocky outcroppings from the right bank (A). Pass just TO THE LEFT SIDE of both of these (they are about 20 feet apart). Now, just in front of you within 30 feet, you will see a rock-ledge (B) running almost parallel to the stream-flow. This midstream partially-submerged ledge is called the 'alligator' because of its spiny-backed appearance. To AVOID THIS 'ALLIGATOR' (B), you must reposition yourself back on the RIGHT SIDE OF THE RIVER. Pass Between rock ledge (C) and right bank. Align yourself with the 'V'-slot (D) which is located about 20 feet in front of you. Note: This 'V'-slot (D) is on the right side, within a FEW FEET OF THE BANK. Be sure your alignment is correct BEFORE ENTERING 'V'-SLOT. BE CAREFUL. Just beyond your entry into the 'V'-slot you will see a submerged boulder (F) on the right side — you must avoid this boulder. ALSO, be careful not to get too far to the left, where another large boulder (E) creates a strong suckhole. Pass through the 'V'-slot in the center of the 2 submerged boulders. This is where the greatest pour-off and hydraulic pressure are concentrated. Resume your position in the center of the river and ride-out the rapids on the standing waves which continue downstream.

Rapid No.	Drift Mileage	Time	Class	Name	
---	5.1	---	---	**WILDCAT STILL-WATER CAMPSITE**	Located on the left bank at the lower end of the bar. Water is found at the downstream end of the bar. This campsite is suitable for large parties, and can serve as a re-grouping site for parties working their way through Wildcat Rapids.
--	5.2	---	---	**RUSSIAN CREEK**	Russian Creek Campsite and Russian Creek are located on the right bank. There is also a campsite, directly across the river, on the left bank - but water must be transported to it.
11	5.2	7 min	II+	**RUSSIAN RAPIDS**	This area is named after a Russian gold-prospector who lived in this area.

HOW TO: Approach by positioning yourself 40% off the **LEFT** bank. Now, the water pressure will try to force you into the right bank, so you must **MAINTAIN POSITION** approximately 40% off the left bank. This may take considerable power. Proceed through the rapids in this position. The first drop is the greatest, and large standing waves continue for a hundred feet or more. Once you enter these standing waves the current tries to push you into the left bank, so you must counteract by positioning yourself in the center of the waves at about their halfway point. Move to slightly **RIGHT OF CENTER** during the passage through the lower section of standing waves. Russian Bar is located on the right as you pass through.

Russian Rapids.

Rapid No.	Mileage	Drift Time	Class	Name	
12	5.5	3 min	II+	UPPER MONT-GOMERY CREEK RAPIDS	Once past Russian Rapids, keep alert so that you can identify several mid-stream boulders which act as a check-point for Montgomery Creek Rapids. The rapids begin approximately 100 yards downstream from these boulders. Stay in the right channel until past the boulders, then move to the left side of the river. During the Depression years there was extensive gold-mining on Montgomery Creek. Over 25 buildings were located on the north bank of the river at that time. The 1955 flood destroyed the buildings.

HOW TO: When past the above-mentioned boulders, move to the extreme **LEFT SIDE OF THE RIVER** (so close to the bank that your craft almost touches it). At the very beginning, a narrow chute between the shore and a boulder can be seen (only **10-12 feet wide**) — you must **PASS THROUGH THE NARROW CHUTE.** Once through, you must immediately position yourself in the **CENTER** of the river to avoid another boulder ahead of you some **20 feet.** The water now broadens and smooths for some 75 yards in preparation for Lower Montgomery Creek Rapids.

Rapid No.	Mileage	Drift Time	Class	Name	
14	5.6	1 min	II+	LOWER MONT-GOMERY CREEK RAPIDS	***SCOUT ON RIGHT BANK!

HOW TO: Position yourself in exact **CENTER** of the stream. Identify the center standing-wave (in low water this can be seen resulting from a sharp spiked rock). There are 2,3 or 4 other large swelling standing waves within a few feet of each other across the entire river width (depends upon depth of water). All of these waves are caused by massive submerged or partly-submerged boulders. You must pass within one or two feet of the spiked rock (or its standing wave) on its **RIGHT SIDE.** To do this you will slip through a narrow chute between the spike and a boulder just 6 or 8 feet to the right of it. This is your **TARGET** — if necessary, check from shore before going farther. The chute you enter has the **LEAST DANGER** from backwash and eddies, but it is not straight downstream — it angles from the center towards the left bank as the water pours off the left side of the submerged boulder. Therefore, you must: (a) avoid the spike by passing it so that it is just a few inches on your left side, while at the same time (b) avoid hitting the boulder which will be a few feet on your right side. Done perfectly, you will not hit either obstruction (especially in low water). Anything other than perfect will result in a hit. **CAUTION:** the rock spike can tear a hole in the bottom of your vessel.

Once past this tricky spot, the stream continues rapidly for a few hundred feet, then broadens and slows. Montgomery Creek enters the river on the left side. This segment of the river once had between 20 to 30 gold-mining operations.

Rapid No.	Mileage	Drift Time	Class	Name	
15	6.1	2½ min	II+	HOWARD CREEK CHUTES	This rapids is made up of a series of drop-offs, each of which forms a chute.
				CHUTE NO. 1	**HOW TO:** About 10% of the river passes to the left of a cluster of rounded boulders. You must take the **MIDDLE CHANNEL.** Stay in the center of the large standing waves.
				CHUTE NO. 2	**HOW TO:** Some **100 yards** downstream from Chute No. 1 (30 seconds), you enter Chute No. 2. Approach and pass through large standing waves in **DEAD-CENTER.** Watch out for rock-spike just left of center.
				CHUTE NO. 3	**HOW TO:** Another **120 yards** (35 seconds) downstream you will see a large boulder in the right center of the stream. **FIRST:** Pass down the main stream in **CENTER** of channel, but **SECOND:** Pull to the right to avoid a downstream submerged boulder (about 60 feet past the large boulder). Once past the submerged boulder, water calms.

Approximately 200 yards downstream on the left bank, Howard Creek enters the river. Use this as a CHECK-POINT . . .

Turkey vultures preen along river's edge.

Boatman negotiates Slim Pickens rapids.

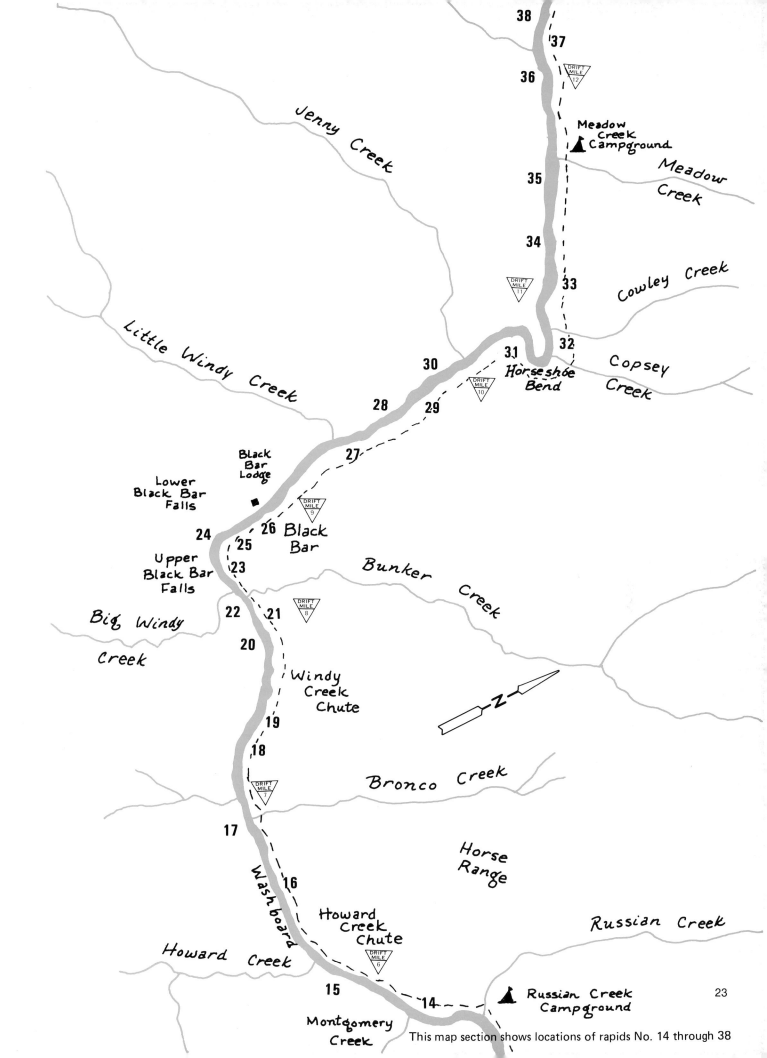

Jenny Creek

Little Windy Creek

Black Bar Lodge

Lower Black Bar Falls

24

26 Black Bar

25

Upper Black Bar Falls

23

Big Windy

22 21

Creek

20

Windy Creek Chute

19

18

17

Washboard

16

Howard Creek Chute

Howard Creek

15

Montgomery Creek

38

37

36

DRIFT MILE 12

Meadow Creek Campground

Meadow Creek

35

34

33

Cowley Creek

DRIFT MILE 11

32

31 Horseshoe Bend

30

Copsey Creek

28 29

27

DRIFT MILE 10

DRIFT MILE 9

Bunker Creek

DRIFT MILE 8

N

Bronco Creek

DRIFT MILE 7

Horse Range

Russian Creek

DRIFT MILE 6

14 Russian Creek Campground

23

This map section shows locations of rapids No. 14 through 38

Rapid No.	Drift Mileage	Time	Class	Name	
***	6.2	---	---	*HOWARD CREEK CHECK-POINT*	*This is your second CHECK-POINT. You can re-assess your trip to this point, regroup your party and relocate your position precisely in this River log.* At this Howard Creek Check-Point there is a small campsite.
16	6.5	9 min	I	**SLATE SLIDE RIFFLE**	This area is named for the slate-shale outcroppings obvious from the trail. The approach to this small rapids can be identified by a massive gray bare-rock outcropping of the mountain wall on the left side of the valley. **HOW TO: Approach this rapids by maintaining your position in MIDDLE OF THE RIVER. Avoid a small rock outcropping from the left bank at the bottom of the riffle.**

Slim Pickens Rapids — note the large metal barge in the foreground.

Rapid No.	Drift Mileage	Time	Class	Name
17	6.9	2 min	III	SLIM PICKENS RAPIDS

*** SCOUT ON RIGHT SIDE

This rapids was named by early boatmen because of the difficulty of passage through its narrow chutes. The chutes were later blown out by dynamite to assure safe and easier passage. Notice on the south bank of the river at the head of the rapids a large steel dredge platform. This dredge was washed downriver from the mouth of Grave Creek by the 1955 flood.

HOW TO: Approach the rapids to the RIGHT OF CENTER-STREAM. Identify a large house-sized boulder approximately 60 yards downstream from the barge. You must pass BETWEEN this large boulder and the right bank. A narrow channel, only 8-10 feet wide is your pathway. Driftboats must pull oars in. Be sure you do not let the current push you into the right bank immediately below the narrow passage. Prevent this by working your way to midstream as soon as you are free of the passageway. Kayaks and rafts may pass to the left of the large boulder, and at levels above 1.5 drift-boats may also consider this left-sided passage.

Slim Pickens Rapids — boatmen must push oars forward through narrowest section.

Swallows' nests on rock outcropping, upstream from Charley's Riffle.

Just below Slim Pickens Rapids a small creek, Bronco Creek, enters the river by working its way through large strewn boulders on the right bank.

The river current slows considerably at this point, causing an increase in drift time.

Rapid No.	Mileage	Drift Time	Class	Name
18	7.1	6 min	I	CHARLEY'S RIFFLE

Approximately 100 yards upstream from this easy whitewater you can see many swallows' nests on the downstream-side of the rock wall on the right side of the river. Also, a sharp, small rock juts up out of the middle of the stream.

HOW TO: Pass to the RIGHT of the jutting rock. Immediately past, set up for Washboard Rapids, which is only some 100 yards downstream.

Rapid No.	Mileage	Drift Time	Class	Name
19	7.2	20 sec	II+	WASH-BOARD RAPIDS

Washboard Rapids consists of 2 narrow chutes, and is located immediately after Charley's Riffle.

HOW TO: CHUTE NO.1 — Position yourself 1/3 OFF LEFT BANK and maintain position.

HOW TO: CHUTE NO.2 — position yourself in EXACT CENTER and pull-off left rock wall. Notice the river channel narrows to its narrowest width on the trip at this point.

Rapid No.	Mileage	Drift Time	Class	Name
20	7.7	4½ min	II+	PLOW-SHARE RAPIDS

This rapids was named because of sharp-edged, plowshare-shaped rocks formed in the bedrock of the right side of the river channel.

HOW TO: Bedrock extends parallel to the river flow. Position yourself in dead CENTER and pass through this small rapids. At high water levels, pull RIGHT at beginning, and keep off the left wall.

Rapid No.	Mileage	Drift Time	Class	Name
21	8.0	10 sec	II	BIG WINDY CHUTE

Only 50 yards downstream from Plowshare Rapids you will enter Big Windy Chute.

HOW TO: Simply remain dead CENTER.

Rapid No.	Mileage	Drift Time	Class	Name
22	8.1	10 sec	II	WINDY CREEK CHUTE

This general area was a source of gold for early miners. After passing through Big Windy Chute you will see Windy Creek Chute within another 60 yards.

HOW TO: Position yourself dead CENTER and ride through.

Rapid No.	Mileage	Drift Time	Class	Name
***	8.2	---	---	BIG WINDY CREEK CAMPSITE CHECK-POINT
23	8.5	2 min	III	UPPER BLACK BAR FALLS

This is your third CHECK-POINT.
Bunker Creek enters on the right side approximately 100 yards past Big Windy Chute. On the left another 50 yards is Big Windy Creek.

In the past, both Bunker Creek and Big Windy Creek were mined for gold. Both sluice-boxes and pans were common gold-mining methods. The large rock just upstream from Windy Creek (on the left side) is a natural diving-platform. This is an excellent spot to stop for lunch or camp a small party. The location is beautiful and small trout can be caught in the creek.

***SCOUT ON RIGHT BANK!

Identify by its sound and the narrowing of the river as it is split by boulder formations.

HOW TO: Stay as close to the right bank as possible all the way through this rapids. Position yourself on the extreme **RIGHT** side of the river. Keep your bow just a couple of feet away from the right bank and avoid being forced into the abruptly perpendicular curving-rock-wall by the powerful currents. Don't get too far off the rock wall because a large boulder near mid-stream must also be avoided. You must pass between the large midstream boulder and the sheer rock wall (located 60 yards downstream). Once past this critical point, the river slows and forms a large pool. **HI-WATER OPTION:** Slip along left side.

Aerial view of Upper Black Bar Falls.

Upper Black Bar Falls — safe passage is down the right side.

Rapid No.	Mileage	Drift Time	Class	Name
24	8.6	1 min	III	LOWER BLACK BAR FALLS

Watch for a cold, fresh spring on the left bank (about halfway to Lower Black Bar Falls). This water is reputed to be the coldest on the Rogue journey, and offers pleasant refreshment on hot days.

Identify easily because it is the next rapids (within 100 yards after Upper Black Bar Falls.

HOW TO: Position yourself in dead CENTER. Identify two submerged boulders slightly to the left of center. Pass within a foot or two on the right side of these two boulders. Remain aligned in CENTER over the standing waves which follow.

Lower Black Bar Falls — lots of whitewater.

| 25 | 8.7 | 2 min | I | BLACK BAR RIFFLE |

Approximately 200 yards downstream from Lower Black Bar Falls.

HOW TO: This minor rapids can be easily negotiated by staying slightly RIGHT OF CENTER. Notice several protruding rocky walls extending from left bank 1/3 into the river, halfway through the riffle.

Rapid No.	Drift Mileage	Time	Class	Name
---	8.8	---	---	BLACK BAR

Black Bar is an alluvial gravel bar about 300 yards wide and 800 yards long. It is located on the left bank of the Rogue. This Bar was named after a man named Black, who was killed here and allegedly put into his boat and shoved into the river by his assailant. His body was found downstream. The lodge was built in 1932 and is currently owned by Bill and Sally Hull. Since the original lodge was built many smaller cabins have been added by Red Keller, who has spent almost 40 years of his life on the banks of the Rogue. The Lodge is open from approximately Memorial Day to mid-November. Reservations are required. Before the 1964 flood, an airstrip was active on the bar, but the flood destroyed it. The flood reached such proportions in this part of the river that the Lodge itself was almost three feet deep with Rogue water. When at the Lodge, be sure to ask Bill, Sally or Red for the story behind the airplane crash and Glen Wooldridge's amazing rescue feat. If you are interested in finding gold, the cracks and crevices along the left bank contain deposits of gold which can still be retrieved with a pan, even today.

Red Keller and dog (on left)

Rapid No.	Mileage	Drift Time	Class	Name
26	8.9	2 min	I	LOWER BLACK BAR RIFFLE

Identify by its noise and location within 200 yards of Black Bar Riffle.

HOW TO: Position yourself in exact CENTER for first 50 yards, then move to LEFT OF CENTER to avoid rock outcropping at lower end of riffle.

| 27 | 9.3 | 5 min | II | LITTLE WINDY CREEK RIFFLE |

A campsite is located on the left side, just above the riffle, suitable for a medium-sized party. There is plenty of water, and a toilet is available. Identify the riffle by locating Little Windy Creek as it enters the river on the left side. 100 yards farther marks the beginning of Little Windy Creek Riffle.

HOW TO: Position yourself slightly LEFT OF CENTER. Remain in this approximate position as you maneuver slightly to avoid large standing waves on both sides.

| 28 | 9.5 | 3 min | I | STANDING WAVE NO. 1 |

Located in center of river. Caused by a submerged boulder. Easily avoided.

HOW TO: Pass to either side. No problem.

| 29 | 9.8 | 4 min | I | RIFFLE NO. 29 |

Identify by locating a rock spike in right center of river (which may appear as a small standing wave in high water). A large boulder breaks the surface within 20 feet of the rock spike and other rocks break surface or cause standing waves throughout the riffle.

HOW TO: Remain in the general CENTER of the river. Maneuver slightly to avoid standing waves and protruding rocks which you will pass on both sides.

Black Bar Lodge.

Aerial view — extends from Black Bar (top) to the Telephone Hole below Horseshoe Bend.

THE RIVER SAUNA requires a large tarp or plastic ground cloth (at least 12 by 14 feet) suspended between rocks, trees or oars as shown above. Tuft-tie golfball sized stones at the corners. To make a tie push the stone from the underside of the tarp, then push the tarp around the sides of the stone and pinch in below the stone (tuft). Now tie the rope around the pinched neck of the tarp, this will trap the stone and tarp so that neither will pull free. Be sure and use smooth stones to prevent punctures of the tarp.

HEAT SOURCE is a metal bucket of hot rocks, which have been gathered above water level and heated until glowing red. Use shovel to transfer rocks from fire to bucket. A separate bucket (could use bailing pail) is used for water source. Participants should wear minimal swimsuit or wrap-around towel. Gradually splash water on hot rocks keeping head low as temperature rises. After each person reaches their desired "sweat" dash out and cool off in the river. Diving head first is not advised unless river depth has been tested prior to the SAUNA.

33

Rapids at Horseshoe Bend can be coordinated with 'HOW TO' below.

Rapid No.	Drift Mileage	Time	Class	Name	
30	9.9	2 min	I	SHADY CREEK RIFFLE	Identify by a rock outcropping in right center of the river. **HOW TO: Pass directly down DEAD CENTER. Maneuver slightly to avoid rocks in midstream.**
***	10.1	---	---	*JENNY CREEK CHECK-POINT*	*This is your fourth CHECK-POINT.* *Jenny Creek enters on the left side. Good campsite.* This was also a popular gold mining area. Considerable mining activities resulted in large quantities on mining equipment being left here, but the flood of 1964 washed much of it away. Miners used a cable-car to cross the river at this point. In November, 1855, an attempt by 386 volunteers and 50 Army regulars to capture rebel Rogue Indians was aborted at this site when the Indians fired heavily upon the soldiers from the south bank. After five or six hours of fighting, the volunteers and soldiers admitted the futility of their efforts and left with one man dead and four wounded.
31	10.3	3 min		HORSE-SHOE BEND RAPIDS	This is a series of 3 stretches of rapids. Numbers 1 and 2 are riffles, Number 3 is the major rapid of the series.
			II	*RIFFLE NO. 1*	Identify Riffle No. 1 because of the development of whitewater after passing through a couple of hundred yards of stillwater. A small boulder bar extends into the river from the right bank. **HOW TO: Position yourself in MIDDLE of stream. Be careful to avoid submerged boulders on left.**
			II	*RIFFLE NO. 2*	You will now enter immediately into Riffle No. 2. **HOW TO: Position yourself in MIDSTREAM and pass through the riffle in this position at all times. Maneuver slightly to avoid minor standing waves and obstructions.**
			III	*RAPIDS NO. 3*	You are now approaching the tough one. Look ahead. Water wants to sweep you into the left bank, with a big submerged boulder in front of a rock-face wall. This is a dangerous boulder and has taken several lives and many boats! Be sure to identify it and avoid it; otherwise, the passage is not too difficult. **HOW TO: Approach by positioning yourself on RIGHT QUARTER OF THE RIVER. As water starts to push you to the left side, you must hold position in the MIDDLE of the stream to avoid hitting the submerged boulder! The river bends and currents try to force you into the boulder and the wall. Hold off and sweep around bend in MIDDLE of stream. Current now veers from bank-to-bank. Just keep your craft positioned in MIDDLE of the river and avoid obstacles.**
32	10.7	4 min	II		Identify at end of Horseshoe Bend, and by the large rock 120 yards before the riffle. The rock stands alone dramatically in the center of the river. The depth at this stretch of the river is almost 70 feet. Good steelhead fishing in colder water temperatures. This is also a likely place for deep-lying sturgeon. Watch for bear along the riverbanks — they frequent this area. Looking back up the river, towards Horseshoe Bend, a magnificent view may be seen. **HOW TO: Position yourself in MIDSTREAM. Be careful to avoid submerged boulders on left. Identify sharp spine protruding from water in midstream and pass BETWEEN it and jagged high rock on right side of river.**

Rapid No.	Mileage	Drift Time	Class	Name	
33	11.1	4 min	II	**TELEPHONE HOLE RIFFLE**	Named because of a Forest Service Telephone line which crossed the river at this point. A good place to eat lunch on a hot day because of shade on the north bank. Identify by a gravel bar on the left, protrusions on the right, a house-sized boulder in mid-stream approximately 100 yards downstream from the beginning of the riffle.

HOW TO: Position yourself on the left 1/3 of the river until approximately 30 yards from the 'house rock', then move to the right 1/3 of the river and pass the 'house rock' in the CENTER of the channel on the RIGHT side of the rock.

Rapid No.	Mileage	Drift Time	Class	Name	
34	11.3	2¾ min	I	**UPPER MEADOW CREEK RIFFLE**	Identify by an obvious narrowing of the river as large boulders restrict the river from the right side.

HOW TO: Approach in midstream. Be sure to stay in LEFT CENTER until The halfway point (into the riffle). Then watch for large submerged boulder against left shore, and avoid by moving to RIGHT CENTER. Another 50 yards brings you to a submerged boulder in midstream, which is easily avoided by passing on its LEFT side.

Rapid No.	Mileage	Drift Time	Class	Name	
35	11.5	2 min	II	**MAIN MEADOW CREEK RIFFLE**	Identify because of a large boulder at its beginning which extends 1/4 of the way into the river from the right bank, and which splits the river. The main channel is on the LEFT side.

HOW TO: Position yourself 1/3 OFF LEFT BANK, between the large rock and the left bank. Maintain position down the center, as the river narrows. Ride out the rapids on the crests of the standing waves which are formed.

Rapid No.	Mileage	Drift Time	Class	Name	
***	---	---	---	*MEADOW CREEK CHECK-POINT*	*This is your fifth CHECK-POINT.*

Meadow Creek was named because of the large meadows on the hillsides surrounding this area. A semi-permanent guide-camp used to be located here before the B.L.M. removed it. This is a very popular campsite and has been used by commercial guides because of the Fall steelhead runs.

There is an old winch, which was used by miners to move the large boulders on the bar, located on the gravel bar along the river, just above the creek. The winch was cranked by hand, and allowed the miners to mine the gold-bearing sands that accumulated around the bases of the boulders.

In April, 1856, General Lamerick and two battalions considered crossing the Rogue to capture rebel Indians near this point. However, he was convinced to reconsider by Major James Bruce, who claimed the medical facilities at Little Meadow would be too far away to adequately handle casualties. This area's forest meadows were evident hundreds of years ago. They became a stronghold for rebel Indians during the Indian Wars of 1855-56.

From this point, on your river journey, be always alert for random midstream boulders which become more frequent downriver.

Rapid No.	Mileage	Drift Time	Class	Name	
36	12.0	8 min	II	**UPPER DULOG RAPIDS**	Identify because of a campsite on the south side of the river, immediately upstream of the rapids. The hiker's trail cuts into the cliff above and ahead. Large house-sized rocks jut up from the banks and riverbed.

HOW TO: Position and stay about 1/3 OFF LEFT BANK, then identify a large rock in midstream. Pass to the IMMEDIATE RIGHT of that large 'house-size' rock (within 1 foot of it).

Rapid No.	Mileage	Drift Time	Class	Name	
37	12.1	2 min	II	**DULOG RAPIDS**	Identify because Dulog Creek enters on the left side, adjacent to Dulog Rapids. Large house-sized boulders split the river.

HOW TO: Position yourself to enter down the CENTER. Stay between the left bank and the large automobile-sized boulder, move closer to the LEFT bank so that you can pass in the channel between several obstructing boulders (which are partially or completely submerged).

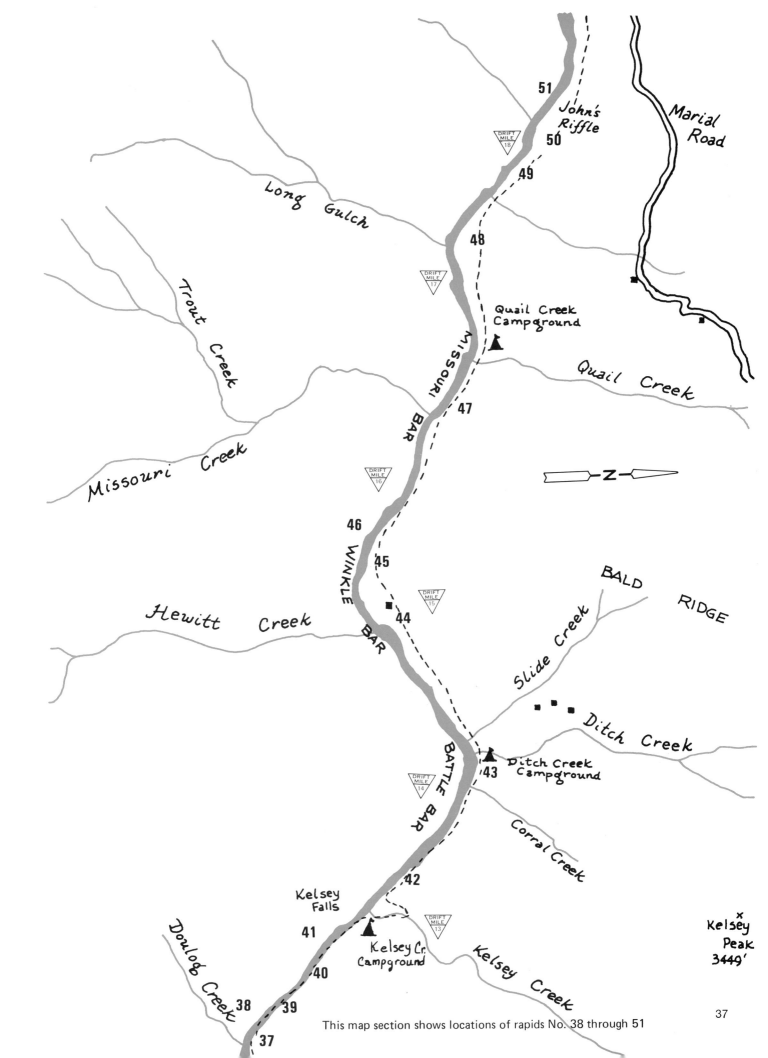

Long Gulch

Trout Creek

Missouri Creek

Hewitt Creek

Doulob Creek

51

John's Riffle

50

49

DRIFT MILE 18

48

DRIFT MILE 17

Quail Creek Campground

47

Marial Road

Quail Creek

-Z-

DRIFT MILE 16

46

45

WINKLE BAR

44

DRIFT MILE 15

BALD RIDGE

Slide Creek

Ditch Creek

43

Ditch Creek Campground

BATTLE BAR

DRIFT MILE 14

Corral Creek

42

Kelsey Falls

41

DRIFT MILE 13

Kelsey Cr. Campground

Kelsey Creek

Kelsey Peak 3449'

40

38

39

37

This map section shows locations of rapids No. 38 through 51

37

Dave Boals and friend.

Black Bear with Chinook Salmon. — Courtesy W. Husum

This osprey nest can be seen near John's Riffle.

Hikers inspect a large Pacific Madrone along the Rogue River Trail.

Entrance to Kelsey Canyon — identify Oodle Rapids, No Name Rapids and Kelsey Falls in succession.

Rapid No.	Mileage	Drift Time	Class	Name	
***	---	---	---	*ENTRANCE TO KELSEY CANYON*	*You are now at the ENTRANCE TO KELSEY CANYON. Note that the canyon walls narrow and become vertical. Hydraulics are strong. Water churns and boils swiftly through this constricted passageway. Average depth of water is 40 feet. Sharp walls and turbulence force boats against sides of canyon. Accidents, injuries and damage occur most frequently in this section.*
38	12.2	1 min	II	**OODLE RAPIDS**	Identify because the river makes a sharp 'S'-turn in this stretch. Also a campsite is located on the south side of the river immediately upstream from the rapids. Looking ahead, you can see the hiker's trail cutting into the cliff above on the right. Large house-sized rocks jut up from the river-banks and riverbed. **HOW TO: Position and stay about 1/3 OFF LEFT BANK, then identify a large rock in midstream. Pass to the IMMEDIATE RIGHT OR LEFT of that large house-sized rock (remaining within 1-foot of the rock).**
39	12.3	1 min	II	**NO NAME RAPIDS**	Identify because it is another 175 yards below Oodle Rapids. A sheer vertical broken wall can be seen immediately after this rapids. **HOW TO: Straight down CENTER.**
40	12.6	3 min	II+	**KELSEY FALLS**	Identify because it is some 500 yards downstream from No Name Rapids. A large, house-sized boulder splits the river. **HOW TO: Take LEFT CHANNEL. Remain in CENTER. Avoid large submerged and semi-submerged boulders.**

Rapid No.	Mileage	Drift Time	Class	Name	
41	12.7	2½ min	II	**LOWER KELSEY FALLS**	Identify because it is some 200 yards downstream from Kelsey Falls. A medium-sized boulder splits the river.

HOW TO: Enter on RIGHT SIDE of boulder. Look ahead 50 feet to drop-off caused by submerged boulder in center of river. Pass this submerged boulder on its IMMEDIATE LEFT SIDE (within inches, if possible). This boulder is not visible at river-levels above 0.8.

***	*13.1*	---	---	*KELSEY CREEK CHECK-POINT*	*This is your sixth CHECK POINT.*

Kelsey Creek enters on the right side, approximately 5-minutes drift-time downstream from Lower Kelsey Falls. A good camp exists here for boaters and hikers. There is good water, toilets and plenty of space.

A short hike up to the trail brings you to a beautiful bridge which permits trail-travelers to cross Kelsey Creek. Kelsey Creek and Kelsey Canyon were named first by Glen Wooldridge in memory of Colonel John Kelsey, led a group of territorial volunteers against Chief John's Indian forces in the spring of 1856. Kelsey Creek is also the present-day boundary between Josephine County and Curry County. During the summer months, salmon may be found schooling at the mouth of this creek.

42	13.4	4 min	I	**LOWER KELSEY RIFFLE**	Identify by white riffles in streamflow and large rock in right center stream.

HOW TO: Pass on LEFT SIDE of large rock in right riverbed. Stay in the CENTER of stream. Avoid small obstacles as necessary.

---	13.5	---	---	STURGEON ROCK	This area has also been given the name "Sports Illustrated Rock" because of a fishing party from Sports Illustrated Magazine, which had excellent Steelhead fishing results in this spot. Water at this section of the Rogue is some 50 feet deep. Hidden behind the large rocks on the left side you can find an excellent campsite. Sturgeon Rock itself is a large rock on the left (south) side of the river.

---	14.0	---	---	BATTLE BAR	Battle bar was the site of a battle between the U.S. Cavalry and a band of Indians in 1856. A cavalry of 536 men, under the command of Colonel Kelsey, attacked an Indian encampment consisting of 200 Indians, mostly women and children, on Battle Bar on April 27. Colonel Kelsey's men would not cross the river in their collapsible boats. They remained on the north bank and exchanged rifle fire with the Indians. Reportedly, 20-30 Indians were killed and 1 soldier died. Finally, the soldiers returned to their home camp without defeating the Indians. Souvenir arrowheads and bullets may still be found with careful searching. Remnants of early gold mining efforts can be identified because many of the holes which were dug still have not been filled in on the gravel bar. On his first trip down the river, Zane Grey camped here and spent over a week hiking and fishing along the banks. Ditch Creek enters on the right side and supplies a campground with good water. Natural wildlife abounds in this area, and there have even been reports of wild turkeys sighted along these banks.

Another interesting bit of history relates the story of Jack Mahoney and Bob Fox. The Bob Fox cabin is visible on the left bank. Fox's cabin was built in the 1920's. The cabin was partially destroyed by the 1964 flood which covered Battle Bar. Jack Mahoney lived downriver about a half-mile. In early May, 1947, Fox allegedly killed Mahoney's pet deer. Mahoney supposedly waylaid Fox in his cabin, and shot him in the stomach. Fox lay in his cabin, wounded with his stomach injury, for several days. He shot several holes through the cabin roof trying to signal for help, but he finally died on May 10. Mahoney, too, was found dead several days later at the lower end of Half Moon Bar — and there is still controversy today about whether he killed himself or not. Another factor in this unusual case is that Mahoney apparently had a 'death list' upon which was named several other intended victims. Glen Wooldridge was about to make his first motor boat trip upriver from Gold Beach at the time, and his name was on the 'death list.' Investigators found the death list, advised postponement of Wooldridge's trip and finally found Mahoney's body, which completed the bizarre incident.

Rapid No.	Mileage	Drift Time	Class	Name	
43	14.2	14 min	II	BATTLE BAR RIFFLE	Identify because of an old cabin on left bank, which can be used as a shelter. This is the Bob Fox cabin mentioned above.

HOW TO: Stay on RIGHT SIDE of the river until large boulders (about 175 yards downstream from beginning of riffle) stand out of river in midstream. Remain in RIGHT CENTER and pass very close on immediate RIGHT SIDE (within a foot) of the large boulders. Immediately after passing these 3 large boulders, reposition yourself at LEFT CENTER to AVOID SHOALS which develop within a few yards in the right river bed. Finally, at the foot of the rapids, pass down the CENTER of medium standing waves (to avoid shallows on left and right).

Rapid No.	Mileage	Drift Time	Class	Name	
44	14.9	7 min	II	MAIN WINKLE BAR RIFFLE	Identify by a large opening in the valley, Hewitt Creek enters river on left. Winkle Bar is located on the right. On the large flat on the left side of the river above Hewitt Creek the broken-down remains of Jack Mahoney's cabin can be found. Mahoney is the man who allegedly killed Bob Fox in May, 1947.

HOW TO: Approach in CENTER of river and maintain position by riding the crests of the standing waves to avoid gravel shoals on both sides.

Winkle Bar.

41

One of Zane Grey's boats is preserved at Winkle Bar.

NO CAMPING
EXTREME FIRE HAZARD
OREGON
SPORTSMAN'S ASSN.

Zane Grey's cabin at Winkle Bar.

Rapid No.	Mileage	Drift Time	Class	Name	
---	15.2	---	---	WINKLE BAR	This is the site of Zane Grey's cabin. He purchased this mining claim from a gold prospector in 1926. Several of his old cabins, and an old wooden river boat used during the 1920's, are still preserved. There is also a new summer cabin built next to the old cabin, an airstrip and a welcome from the present-day owners as long as you 'take nothing but pictures, and leave nothing but footprints.' Zane Grey's heirs have recently sold Winkle Bar. This area is a very likely place to see deer, and this is one of the most well-known fishing riffles on the Rogue. Across from Winkle Bar, Hewitt Creek campsite is an excellent spot to have lunch, with room for a small camping party.
45	15.5	3¾ min	I	LOWER WINKLE BAR RIFFLE	Identify by its location approximately 200 yards past the lower end of Winkle Bar. This Lower Winkle Bar Riffle consists of TWO PARTS, some 75 yards apart. **HOW TO: FIRST PART: Approach on the right side of the river. A protruding rock splits the river. Take RIGHT CHANNEL. Stay in MIDDLE of channel. Pass on through a 75 yard calm stretch to the SECOND PART: The river shallows here, so remain in dead-CENTER on the tallest standing wave crests.**
46	15.7	2 min	II	ZANE GREY RIFFLE	Identify by a gravel bar extending into the river from the right bank, causing the river to make a curving course in a clockwise direction around it. Several rocks jut out of the left riverbed downstream in the whitewater. **HOW TO: Remain in MIDDLE of the river, but maneuver slightly to barely avoid large protruding rocks which jut out of the river on the right side. Pass obstructions on their left side, as closely as possible. Approximately 75 yards into the riffle, move to RIGHT CENTER to avoid rock jutting up from left riverbed.**
***	16.3	---	---	MISSOURI CREEK CHECK-POINT	*This is your seventh CHECK-POINT. Missouri Creek Checkpoint is located some 13 minutes of drift time downstream from Zane Grey Riffle. Missouri Creek enters the Rogue on its left (south) side. Just below this confluence can be seen Missouri Bar.* Gerald Fry is an Indian who lives in the cabin located on the upper end of Missouri Bar. Fry cares for the buildings at Winkle Bar. You may see his jet boat along the bank somewhere between Winkle Bar and Missouri Bar. Missouri Creek marks the site of heavy gold-mining in early days.
47	16.4	40 sec	II	MISSOURI CREEK RIFFLE	Identify by Missouri Creek entering the Rogue on the left side, some 60 yards before the riffle begins. **HOW TO: Enter the riffle by positioning yourself in dead CENTER and then hold your centered position as you pass through relatively simple whitewater.**
---	16.6	---	---	QUAIL CREEK	This area marks the site of the 1970 Quail Creek fire. The fire burned 2700 acres of forest and grassland on both sides of the river. It was man-caused. Two firefighters lost their lives, many animals were destroyed. The area is still recuperating from the fire. There is a campsite above the hiker's trail.
48	17.3	14 min	II	LONG GULCH CREEK RAPIDS	Identify by Long Gulch Creek entering the river on the left side, some 150 yards before the beginning of the rapids. On the high bank, on the left side, by the creek, several cabins were built by Glen Wooldridge. The Wild River Act resulted in their destruction in the 1970's. **HOW TO: Identify a large rock in the left center stream. Pass close by the rock on its right side and remain in the CENTER of the river for the next 150 yards of calm water. Now you will enter the major part of the rapids. Position yourself in the exact CENTER of the stream. Jutting rocks will pass close by on your left hand side. Downstream, identify a single automobile-sized rock jutting out of the middle of the stream. Pass it CLOSE BY (within 1 foot) by passing it just to ITS LEFT SIDE. Another 60 yards and you will pass on the left side of a massive rock about 2 stories high. And within another 50 feet, pass on the left side of an automobile-sized boulder jutting out of midstream. Finally, another building-sized boulder on your right side marks the end of the rapids. You can pass this final boulder in mid-stream.**

Aerial view extends from Long Gulch (foreground) to John's Rapids (in distance). Note dead vegetation remains from Quail Creek fire in 1970.

Rapid No.	Mileage	Drift Time	Class	Name	
49	17.8	3 min	II	**BIG BOULDER RAPIDS**	Identify by a huge monolithic 3-story-tall boulder on the left side of the river, which marks the position of these rapids. Landslides in this area have caused these boulders.
					HOW TO: Position yourself in RIGHT CENTER of the stream and avoid rock outcroppings adjacent to the huge boulder on your left side. Move to left center immediately AFTER passing the huge boulder, to avoid rock jutting out of the right riverbed.
50	18.1	2 min	II	**ISLAND RAPIDS**	Identify by large boulders lying strewn randomly around the riverbed. A gravel island splits the river ahead. Evidence of the Quail Creek fire is on both sides of the river. Slides are visible on the north bank. Main channel of river is on LEFT side of island ahead.
					HOW TO: Position yourself in MIDSTREAM in the LEFT CHANNEL. Avoid minor obstructions as you hold position through the rapids.
---	18.2	---	---	UNIM-PROVED CAMPSITE	Within 50 yards of passing through Island Rapids, look for a small, un-named creek entering the Rogue on its left side. This fresh-water supply can be used if you choose to camp on the right side of the river. No improvements have been made here.
51	18.5	4 min	II+	**JOHN'S RAPIDS**	Named after Chief John, leader of the Indian tribes during the uprisings of 1855-56. Identify by a small gravel bar on left, about 60 yards before whitewater. Large boulders on left bank indicate beginning of riffle.
					HOW TO: Approach down CENTER OF MAIN CHUTE, rock outcroppings exist on both sides. Pass very closely-by rock on right side of river to obtain easiest passage.
52	18.9	4 min	II	**MAGGIE'S RIFFLE**	Named by Glen Wooldridge after Maggie Stoddard, who caught her first steelhead here. Identify by the boulders from a slide on the right bank, for some 50 yards to the entrance.
					HOW TO: River narrows to swift chute on left side. Position yourself slightly RIGHT OF CENTER of the chute. Adjust to maintain this position through the rapids.
53	19.0	1 min	II	**CHINA RIFFLE**	Naming of this area refers to the communities of Chinese miners living here in early days. Identify by large, monolithic boulders some 100 yards below Maggie's Riffle.
					HOW TO: Position yourself just slightly on the LEFT OF DEAD-CENTER and hold position.
54	19.1	30 sec	II	**CHINA RAPIDS**	Identify by the large monolithic boulders and a distance of some 50 yards from China Riffle.
					HOW TO: Start in DEAD CENTER OF CHUTE, move TO RIGHT to avoid large standing wave (which is caused by a submerged boulder) at the end of the chute.
55	19.3	4 min	I	**RAPIDS NO. 55**	Identify by a gravel bar on left and low, smooth, rock outcroppings on right some 100 yards before the rapids.
					HOW TO: Position yourself in dead CENTER and ride through on the standing waves of the main chute.
56	19.4	3 min	II	**CHINAMEN RAPIDS**	Identify by the river disappearing ahead, large boulders from an earth-scar on the right, and a large dead fir snag on upper right bank some 50 feet before the rapids. In this area a foot bridge was used to cross the Rogue — legend says that white miners used to throw Chinese miners off the bridge to exact punishment.
					HOW TO: Identify sharp spiked rocks or their standing waves at the entrance to rapids. Position yourself 1/3 off left bank. Enter in CENTER OF LEFT CHUTE. Pass along left bank by remaining within 10 feet of it throughout entire rapids, thereby avoiding currents and obstructions in mid-stream.

John's Rapids.

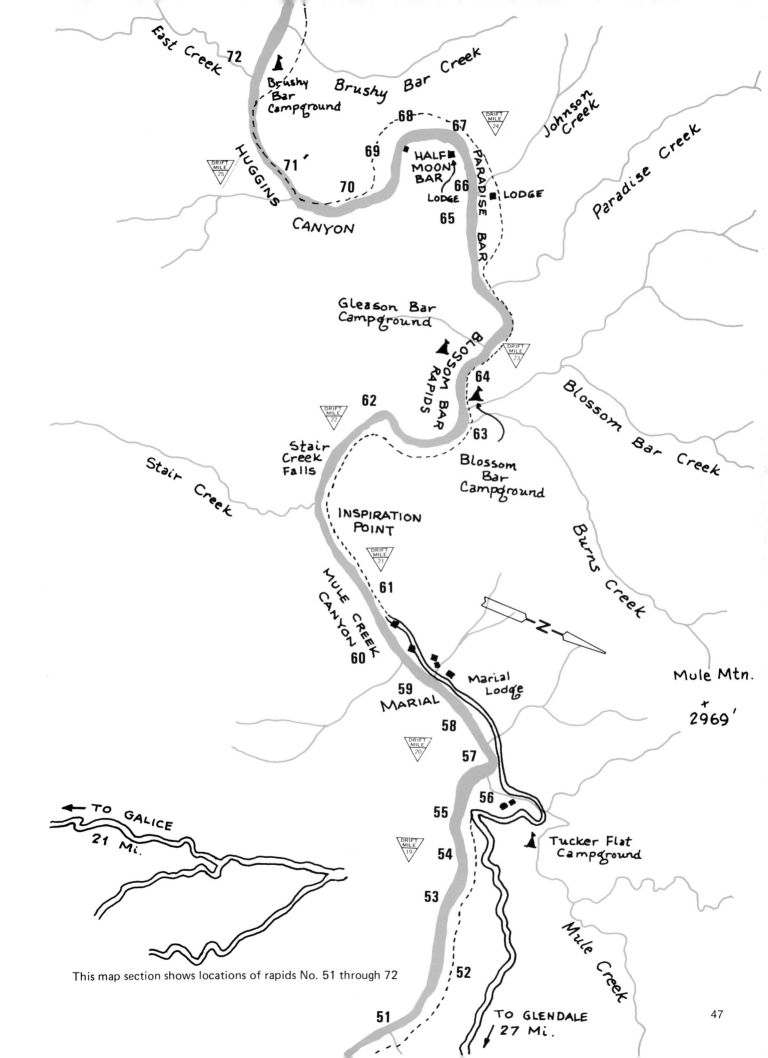

East Creek 72

Brushy Bar Campground

Brushy Bar Creek

Johnson Creek

DRIFT MILE 24

68

67

69

HALF MOON BAR

66

LODGE

LODGE

Paradise Creek

71'

HUGGINS

DRIFT MILE 25

70

65

CANYON

PARADISE BAR

Gleason Bar Campground

BLOSSOM BAR RAPIDS

DRIFT MILE 23

Blossom Bar Creek

64

62

DRIFT MILE 22

63

Blossom Bar Campground

Stair Creek Falls

Stair Creek

INSPIRATION POINT

DRIFT MILE 21

Burns Creek

61

Z

MULE CREEK CANYON

Mule Mtn.

+ 2969'

60

59

58

Marial Lodge

MARIAL

57

DRIFT MILE 20

56

←— TO GALICE
21 Mi.

55

Tucker Flat Campground

54

DRIFT MILE 19

53

52

This map section shows locations of rapids No. 51 through 72

Mule Creek

51

TO GLENDALE
27 Mi.

47

Mule Creek, Mule Creek Eddy and Wild River Ranch.

Rapid No.	Mileage	Drift Time	Class	Name
---	19.6	---	---	MULE CREEK CHECK-POINT

This is your eighth CHECK-POINT. Mule Creek Checkpoint is some 5 minutes drift-time from Chinamen Rapids. A campground is located on the right bank, offering water from the creek. The campground is suitable for large parties.

Mule Creek was named in the summer of 1852 when a company of soldiers from Fort Orford tried to open a trail along the Rogue River in this area. One of the men in the company (named Williamson) rode a mule named John. When the mule was turned loose to graze at the spring or this creek, it wandered off and was not found. This incident resulted in the stream being called 'John Mule Creek', but the name was later shortened to the present 'Mule Creek'. There is also a 'Mule Mountain' nearby. The tale goes on to say that two years later Williamson found his mule at Siletz. This is possibly true, because after the Indian wars in this area, all the local Indians and their possessions were moved up the coast to the Siletz Indian Reservation. This site also marks 'Anderson's Ranch'. The B.L.M. purchased this property in 1970 under the Wild & Scenic Rivers Act. This area was settled by a man named Billings, who homesteaded 70 acres in 1887. If you walk up the creek a short distance you will see a two-story building used by Billings as a General Mercantile Store years ago. The first motor boat to make it up Mule Creek Canyon was piloted bu Ruell Hawkins, in September, 1927. Hawkins had to portage around Blossom Bar, and then continued upstream to Winkle Bar, where Zane Grey's party was camped. Hawkins later was drowned in the Rogue because of a boating accident.

Rapid No.	Drift Mileage	Time	Class	Name
---	19.7	3 min	---	MULE CREEK EDDY
57	19..8	1 min	I	**MULE CREEK RIFFLE**
---	20.1	---	---	MARIAL & MARIAL LODGE

This is a very deep section of the Rogue. We have recorded depths of over 90 feet. Sturgeon lurk in these depths. The greatest depth is located next to the rock ledge on the right. Note the bleached log high on the right bank — it was left there by previous high water. A dry campground is located on the left bank. Rocky Bar Creek enters the river on the right side, just downstream from the rock ledge. This was the site of the last active gold mine on the Rogue River.

This riffle begins approximately 100 feet past Mule Creek Eddy. Marial Lodge can be seen about 400 yards downstream. Notice an old hydraulic water pipe used for hydraulic mining protrudes from the gravel bar on the right.

HOW TO: Position yourself 20 feet OFF LEFT BANK and pass through this riffle in this position. Once through, regain midstream.

Located just a couple of minutes of drift time downstream from Mule Creek Riffle. Before the turn of the century, Tom Billings had a daughter named Marial. Tom and his large family lived on a small ranch and had a few sheep and livestock. Their ranch was located 3 miles up on the mountainside from the river. This was the site of the original Marial Post Office, which was operated by George Washington Billings. The settlement of Marial was later moved to the present riverside location. During the depression, over 250 people received their mail from the Marial Post Office. At one time there were 3 'George Washington' Postmasters in this immediate area, they were: George Washington Billings at Marial, George Washington Missouri at Illahee and George Washington Rialeah at Agness. The first mining in this region was done by John Billings, in 1890. A public access road reaches the Rogue at this point. A small pioneer cemetery is also located here. A campsite is located along the riverbank. Marial Lodge is available for food and lodging if reservations are made in advance. River travelers can walk along the foot-trail at this point to scout Mule Creek Canyon and prepare for the next step of the journey. A walk along the trail (slightly over a mile in length) will bring you to Inspiration Point, overlooking Stair Creek Falls as Stair Creek enters the Rogue on the south side of the river. During the 1964 flood, the river water at this point reached to within a few feet of the lodge. The lodge is run by Ted Camp, one of the most delightful individuals on the river. Watch him, he's a jokester.

Marial Lodge.

Rapid No.	Mileage	Drift Time	Class	Name
58	20.1	2+ min	II	**MARIAL RIFFLE**

Located directly in front of the lodge.

HOW TO: Position yourself 30 feet **OFF LEFT BANK** (current tends to push you into left bank in mid-riffle). Maintain position through the riffle, even though current tries to drive you into left wall.

Rapid No.	Mileage	Drift Time	Class	Name
59	20.4	2½ min	II	**RAPID NO. 59**

Identify because the river takes a bend to the right during this stretch of whitewater. Rock outcroppings extend into the river from the right bank.

HOW TO: Position yourself in **MIDSTREAM** for the **FIRST 50 FEET** . . . then identify 2 standing waves a few feet ahead. **PASS BETWEEN** the submerged boulders causing those standing waves, holding **CLOSE TO THE RIGHT BOULDER.**

Rapid No.	Mileage	Drift Time	Class	Name
60	20.6	1 min	II	**RAPID NO. 60**

Identify by large rocks on right bank, and large boulder causing standing wave in mid-stream.

HOW TO: Position yourself 20 feet **OFF RIGHT BANK.** Hold this position as you maneuver to avoid any obstacles in mid-stream.

Before entering Mule Creek Canyon it is essential that you pull into one of the back-eddies on the left side of the river. Make sure that all gear in your boat is securely fastened, to avoid changes of weight when passing through the canyon. It is also important to make sure that all of your passengers have their life jackets securely in place. If your boat overturns in this section of the river, it would be almost impossible for you or your passengers to get out of the water because of the steep canyon walls. Most of the drownings that have occurred on this section of the river resulted from people failing to wear their life jackets. Loose life jackets have been torn off by the swift rapids.

Entrance to Mule Creek Canyon — 'the jaws'

This view of Mule Creek canyon can be coordinated
with the 'HOW TO'

Rapid No.	Drift Mileage	Time	Class	Name
61	20.8	30 sec	IV	**ENTRANCE TO MULE CREEK CANYON & MULE CREEK CANYON**

Identify because the canyon walls close. Two large monolith boulders mark the entrance (riverguides call it the 'jaws') to the first whitewater. These are named the 'Guardian Rocks' (guaranteed to stimulate chilling thoughts as you drop into this incredible, tomb-like rocky canyon!).

This is a series of 5 whitewater sections which will test your boating abilities!

SECTION 1: ENTRANCE TO CANYON

Identify because the canyon walls close. Two large monolith boulders mark the entrance to this first whitewater.

HOW TO: Approach by positioning yourself just to the **LEFT** of the midstream monolith boulder. Be sure not to go too far to the left, because a gravel bar causes shallows on the left half of the river. Ride on the wake, within a few feet of the monolith boulder, then counteract current's tendency to force you into the left wall of the canyon straight ahead. Once past this dangerous point look ahead immediately to position yourself for the next section, Section 2 . . .

SECTION 2: Located 70 yards downstream from Section 1.

HOW TO: The current crosses back and now tries to force you into the right rock wall. **STRONG HYDRAULICS!** Use power to **KEEP OFF RIGHT WALL** as you are carried past this dangerous section, look ahead immediately, so you can position yourself to navigate Section 3 . . .

SECTION 3: Located 50 yards downstream from Section 2.

HOW TO: Watch out for the partially-submerged boulder on the left wall (riverguides call this 'Telfer's Rock' — named after 3 people who lost their lives here. None of the victims wore life jackets). **DANGEROUS ROCK AND POUR-OFF!** To avoid it, use power to **KEEP AWAY FROM LEFT WALL AND CREST OF THE POUR-OFF** caused by this partially-submerged rock. Once past, look ahead to position yourself for Section 4.

SECTION 4: Located 100 yards downstream from Section 3.

HOW TO: This section is called 'The Narrows' because the canyon restricts severely at this point. You must position yourself in dead **CENTER** as you are swept through this narrow space. You must **PUSH OARS FORWARD TO ALLOW ROOM FOR NARROW PASSAGE. DO NOT SNAG OARS** in this confined space to avoid being tipped over or injured. Once past this short passage, look ahead to position yourself for Section 5. . .

SECTION 5: Located 250 yards downstream from Section 4, around the bend ahead. IMPORTANT: Only one boat at a time through this section! All others hold back in the back-eddy about 50 yards above the Coffeepot, on the left side of the canyon. Be sure the boat in front of you is safely through before entering the Coffeepot.

HOW TO: This section is called 'The Coffeepot' because of the current's welling turbulence which resembles the action of boiling water. At the entrance to this section, the current tries to push you into the right wall, and once into the turbulence, the action of the currents is random and unpredictable. Sometimes you will get through easily, other times it may take twenty minutes to get out of the confined space ahead. Look for stable water to find controlled action for your paddles/oars. It is important to keep your general position **PARALLEL TO THE WALLS** to **PREVENT WEDGING SIDEWAYS** (especially in craft over **15** feet in length). Do not permit passengers to place hands or fingers over edges of boats (avoid amputations!). This is the narrowest passage on the river, so you must handle oars correctly. **PULL IN OARS** to prevent tipping over or injuries. Once past these narrows, the canyon widens and currents stabilize.

Coffeepot in Mule Creek Canyon.

Aerial view of Mule Creek Canyon below the 'Coffeepot'.

Stair Creek Falls.

| --- 21.4 varies --- | STAIR CREEK FALLS CHECK-POINT | This is your ninth CHECK-POINT. Stair Creek Falls Check-Point is a varying drift-time from the Entrance to Mule Creek Canyon because of the unpredictability of your passage through Mule Creek Canyon. Sometimes river-craft pass through with no delays whatever, and only several minutes elapse during their passage to Stair Creek Falls Check-Point. Other times, the river craft are caught in swirling back-eddies and counter currents which do not release their hold for 10 or more minutes. |

Stair Creek cascades into the Rogue River from the south. The stream flows over nearly vertical rock walls to meet the river. Inspiration Point overlooks this scenic spot from the north bank. During summer months this is a likely spot to find salmon refreshing themselves in the cool waters of Stair Creek. An attractive small waterfall is formed by Stair Creek as it cascades into the Rogue. The more hardy adventurer may wish to journey (i.e. climb) several hundred yards up Stair Creek where he will find a small-scale 'Mule Creek Canyon' carved out by the creek.

| 62 | 22.1 | 16 min | I | OLD MINE RIFFLE | Identify because the river sweeps around a bend to the right, and a gravel bar extends into the river from the right bank. Named because of an old mine which is on the south riverbank a few hundred feet before the riffle. |

HOW TO: Pass through in the CENTER of the riffle. Counteract current's tendency to push you into the left wall.

An old gold mine located on the
south side of the river, just above
Rapid No. 62.

Rapid No.	Drift Mileage	Time	Class	Name
63	22.7	4 min	IV	**BLOSSOM BAR RAPIDS**

***SCOUTING MANDATORY *** SCOUT FROM NORTH BANK!

This rapids was named after the wild azaleas that bloom in this area in the spring. Blossom Bar is located on the north bank, and was the site of extensive gold mining operations in the past. Where the hiker's trail crosses Burns Creek, various heavy mining equipment is scattered about. Blossom Bar Creek provides plenty of good water and there is a campsite with out-house facilities. This camp is primarily used by the hikers. In earlier days this rapids was impassable and it took as much as half-a-day to portage around the rapids at this site. Very few boatmen attempted the trip you are now able to make because of the difficulties of portaging the various obstacles and rapids. Glen Wooldridge changed this situation as he cleared the river to make it passable. Wooldridge experimented with several techniques until he developed a method for 'blowing-out' the rocks and boulders. He used a 40% stumping-powder mix for the job. By placing his dynamite into a sack weighted with rocks, he was ready. He would row up to the rock he had chosen to move out of the way, light the dynamite, drop the weighted sack and dynamite overboard on one side of the rock, and then row quickly downstream before the explosion. This area harbors many rattlesnakes and poison oak, so use caution when scouting the rapids. This difficult water stops all upriver motorized boat traffic. This was the site of a large Indian encampment for many years.

HOW TO: As you approach the rapids, approach on the **FAR LEFT** side of the river (under the high rock wall). This channel lasts about 30 yards. Now you must reposition your craft nearer the center of the river to pass near a large boulder overhang in mid-river ('A'). You are now in the section of the rapids known as 'THE HORN' — which is the most dangerous. As you **ENTER THE HORN** (all the currents join together here) **WITH YOUR DRIFT BOAT/RAFT STERN POINTED TOWARDS THE LEFT BANK ('B'),** powerfully **BACK-FERRY TOWARDS THE LEFT BANK** to avoid the large boulders only 20 yards ahead at the bottom of the chute of the 'horn' ('C'). Pass these large boulders at the bottom of the chute of the 'horn' by staying close — **WITHIN 2 FEET ON THEIR LEFT SIDE.** Now you are in position to move towards river-center to line up for passage between large boulders another 30 feet downstream ('D'). As you pass through these final boulders in midstream, begin **PULLING BACK TOWARDS THE LEFT BANK TO AVOID** more randomly positioned and easily avoided boulders for the remaining 200 yards of the rapids.

Dr. David Ross "Doing It Right" at Blossom Bar

Approach on far left

Entering the horn (A)

Boat stern pointed to ward left bank (B)

Avoid large boulder at bottom of the horn (C)

Pass to left of boulders

Pass between boulder down stream (D)

Blossom Bar today. Note several identifiable rock still remain.

Blossom Bar — before Glen Wooldridge dynamited a passageway through the rapids.

Devils Stairs Rapids. Note 3 sections followed by Gleason Creek Campsite on left bank.

Rapid No.	Mileage	Drift Time	Class	Name	
64	22.9	2½ min	III	**DEVILS STAIRS RAPIDS**	The Stillwater section below Blossom Bar gives you a chance to rest. At the end of the quiet water you will easily identify Devils Stairs Rapids. Other names have also been given to this series of pour-offs, which drops 30 feet over a distance of 300 yards. This section has also been called the Devil's Backbone, Devil's Staircase and Devil's Stairstep.

The series is easily divided into three sections:

SECTION 1: HOW TO: A chute identifies the approach. Center yourself in MIDSTREAM and enter chute slot (or 'V'-slot); pass through on center of standing waves. Now look ahead some 30 yards to prepare for Section 2.

SECTION 2: HOW TO: The current forces you into a sheer wall ahead on your right. You must BACK-FERRY STRONGLY INTO MIDSTREAM to keep from being forced into the wall as the current sweeps you past. Now look ahead 75 yards for Section 3.

SECTION 3: HOW TO: Enter in the CENTER and prepare to keep away from the right wall as you reach the end of this section.

---	23.1	---	---	GLEASON CREEK	Gleason Creek enters the Rogue from the south bank. The campground at this location has the capacity for approximately 25 people. On one of our trips down the Rogue, a dog which accompanied us was killed by a rattlesnake bite at this campsite.
---	23.4	---	---	*PARADISE CREEK CHECK-POINT*	*This is your tenth CHECK-POINT. Paradise Creek Check-Point is approximately 6 minutes drift time from Devils Stairs Rapids.*

Paradise Creek cascades over a rock wall into the Rogue from the north. Some years there are sand deposits at this site, permitting camping . . . other years the sand has been washed away.

Breakfast at Gleason Creek campsite.

Rapid No.	Mileage	Drift Time	Class	Name	
65	23.5	5 min	II	**UPPER PARADISE RIFFLE**	Identify by a boulder-bar extending into the river from the right bank; large boulders mark the entrance to the riffle on the left bank; a lodge (Paradise Lodge) marks the end of the riffle ahead on the right bank. **HOW TO: Position yourself in MID-RIFFLE and avoid standing waves as you pass through.**
---	23.6	---	---	PARADISE LODGE	Paradise Lodge is a commercial lodge open year-round. Jet-powered Mail Boats often bring their passengers upstream to this lodge where they lunch and then return to the coast. A 1550-ft. airstrip is located here. You may obtain meals and groceries here, if you wish to pay the necessary premium.

Paradise Lodge.

Rapid No.	Mileage	Drift Time	Class	Name	
66	23.7	1½ min	I+	**LOWER PARADISE RIFFLE**	Identify because it is located 100 yards downstream from the Paradise Lodge beach. **HOW TO: Approach this riffle approximately 1/3 OFF RIGHT BANK and hold position all the way through. Avoid minor obstacles as necessary.**
---	23.9	---	---	HALF MOON BAR	Half Moon Bar is located on the left bank. It has an airstrip. Several privately-owned cabins are located here. Also, a privately-owned lodge is operated here, but is not visible from the river. Reservations are required for lodging at this lodge. This bar was the site of Indian camps for many years.
67	23.9	4 min	II	**UPPER HALF MOON BAR RIFFLE**	Identify by a cable crossing the river directly above the beginning of this riffle. **HOW TO: Position yourself in DEAD CENTER and maintain this position through the riffle.**

Half Moon Bar. Note airstrip and Lodge on bar. The Lodge is not visible from river.

Rapid No.	Mileage	Drift Time	Class	Name	
68	24.3	2½ min	II	MAIN HALF MOON BAR RIFFLE	Identify by a gravel bar extending into the river from the right bank, dramatic sheared away scar marks the rock face of the right cliff. **HOW TO: Position yourself approximately 40% OFF LEFT BANK and maintain this position through the riffle by back-ferrying as necessary.**
69	24.5	1 min	II	LOWER HALF MOON BAR RIFFLE	Identify by a gravel bar extending into river from left. **HOW TO: Position yourself dead CENTER and maintain position by keeping off right side rocks and wall as current tends to move you into them. Now look ahead to the start of Huggins Canyon and Huggins Canyon Rapids series. . .**
---	24.7	½ min	---	HUGGINS CANYON	This section of the Rogue was named by Glen Wooldridge after a local hunter, Andy Huggins, who lived for many years at Half Moon Bar. Huggins' grave is presently located at Half Moon Bar.
70	24.7	½ min	II	HUGGINS CANYON RAPIDS SECTION 1	**HOW TO: Position yourself in LEFT CENTER and allow current to carry you near right bank (near a boulder bar), then stay in MIDSTREAM until the end of the boulder bar. Look ahead 30 yards to position for Section 2.**
			II	RAPIDS SECTION 2	HOW TO: Position yourself in dead CENTER and pass through on standing waves. Look ahead 60 yards to position yourself for Section 3 . . .
			II	RAPIDS SECTION 3	HOW TO: Current tries to push you into left bank, so use power to remain in MIDSTREAM and maintain this position as you pass through this section.

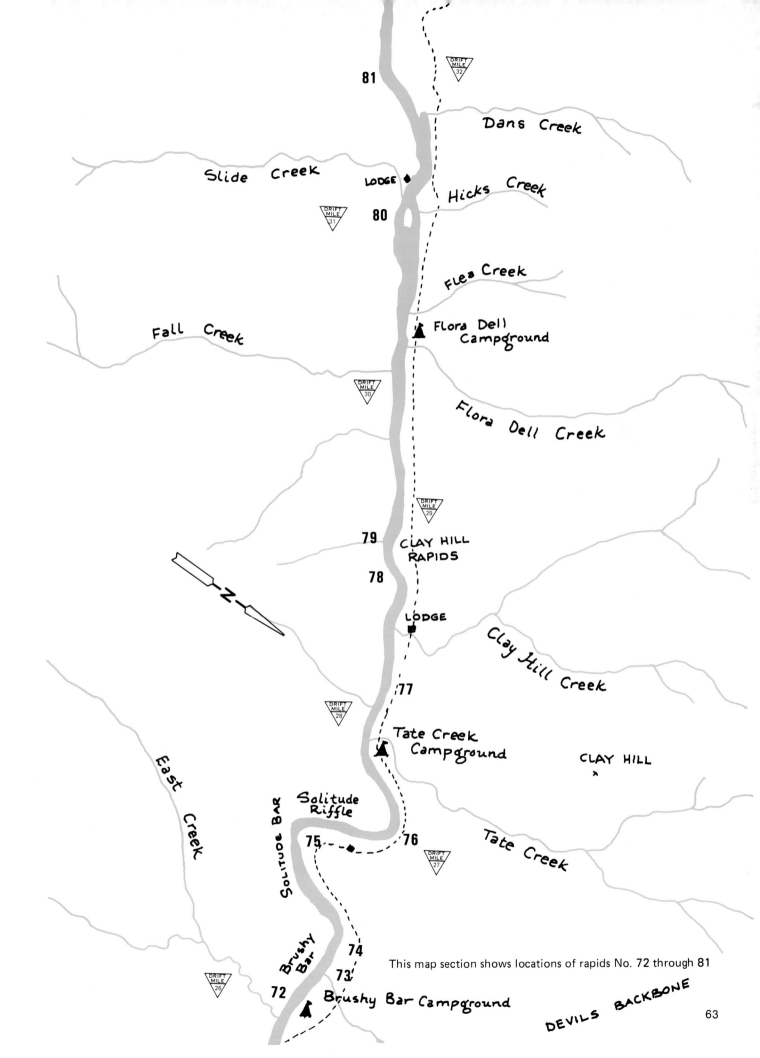

81

DRIFT MILE 32

Dans Creek

Slide Creek

LODGE

80

DRIFT MILE 31

Hicks Creek

Flea Creek

Flora Dell Campground

Fall Creek

DRIFT MILE 30

Flora Dell Creek

DRIFT MILE 29

79

CLAY HILL RAPIDS

78

N

LODGE

Clay Hill Creek

77

DRIFT MILE 28

Tate Creek Campground

CLAY HILL

East Creek

Solitude Bar

Solitude Riffle

75

76

DRIFT MILE 27

Tate Creek

Brushy Bar

74

This map section shows locations of rapids No. 72 through 81

DRIFT MILE 26

73

72

Brushy Bar Campground

DEVILS BACKBONE

63

Rapid No.	Mileage	Drift Time	Class	Name	
71	25.0	5 min	I	**HUGGINS CANYON CHUTE**	Identify by large boulders which appear to have slipped into the river from the right bank. Other boulders match these on the left side and form the chute.
					HOW TO: Position yourself to pass through the CENTER of the 'V'-slot formed between the rocks.
---	25.7	---	---	STURGEON HOLE	This is a deep spot in the river, approximately 70 feet in depth. It is located some 17 minutes drift time from Huggins Canyon Chute. A small creek enters on the left side, forming a gentle waterfall.
---	25.8	---	---	*EAST CREEK CHECK-POINT*	*This is your eleventh CHECK-POINT. East Creek Check-Point is approximately 20 minutes drift time downstream from Huggins Canyon Chute.*
					East Creek enters on the left side. This is the site of the former 'Generals' Cabin', owned by Generals Eaker, Spatz, LeMay, Anderson and Twining. This land was sold to the group by Glen Wooldridge as a former mining claim. The cabin was torn down after the passage of the Wild Rivers Act. The steps up to the cabin are still visible on the steep rock wall, and a fireplace identifies the site of the original building.
72	26.0	2½ min	II	**BRUSHY BAR RIFFLE**	Identify by a low boulder bar entering the river from the left side (its restriction causes the riffle).
					HOW TO: Position yourself in dead CENTER of the standing waves and maintain position on their crests to the end of the riffle.
---	26.0	---	---	BRUSHY BAR	Brushy Bar is not to be confused with the gravel bar on the left. Brushy Bar itself is located on the right bank. Forest Fires burned this area in 1905, resulting in low, dense brush growing over this area, giving it its name. Mining was extensive here and ditches are still evident. A large campsite is available.
73	26.2	4 min	II	**LOWER BRUSHY BAR RIFFLE**	This riffle is located at the lower end of the boulder bar on the left side. Some riverguides call this Black Bear Riffle, and the boulder bar Black Bear Bar. This name is a holdover from the early days' Black Bear Mine which was located at this site.
					HOW TO: Position yourself about 1/3 FROM LEFT BANK to avoid shallows on the right 1/3 of the riverbed. Ride through with currents which carry you to right center and out the bottom end.
74	26.3	1 min	II	**RIFFLE NO. 72**	A gravel bar extends into the river from the right, restricting the river to form this riffle.
					HOW TO: Position yourself about 20 feet off left bank and ride through on the standing waves, holding position by moving to midstream as necessary.
75	26.8	4 min	II	**SOLITUDE RIFFLE**	At this point the river narrows. Ahead, a forest-covered wall can be seen crossing the river as the river bends to the right. At the head of Solitude Bar, on the north side of the river, note the steepness of the hillside above the trail. This area marks the site of Captain William Tichenor's defeat. During the Indian wars of 1855-56, Tichenor and his men were enroute to Illahee to rescue settlers from an Indian attack. At this location, the Indians rolled rocks down the steep hillside on top of the troops, disbanding them. This successful ambush only temporarily postponed Captain Tichenor who, after the Indian wars, helped round up and force all Rogue River Indians onto the Siletz Reservation. Solitude Bar also bustled with mining activities about 1900. The large piles of rocks left from these operations can be found above the trail. A large arrastra wheel is lying in the brush on the left bank at this site. An arrastra was sometimes used by early mining operation to break up the ore to more readily extract the minerals. Watch out for rattlesnakes if you search for the arrastra. A small campsite can be found on the right bank, but it has no water.

Aerial view of the Rogue River canyon, from Paradise Bar (right) to Brushy Bar (left).

Female Merganser followed by her family.

Rainie Falls Reject.

Raft and Drift Boat Sandwich at Blossom Bar.

Rapid No.	Mileage	Drift Time	Class	Name	
75 (continued)	26.8	4 min	II	SOLITUDE RIFFLE	**HOW TO: Identify the 'V'-slot (located approximately 30 feet from the right bank). Enter the 'V' and ride the crests of standing waves approximately 400 yards downstream to pass between 2 large monolithic rocks at the end of the standing waves. After rounding the bend in the river, an easily negotiated chute develops some 100 yards further downstream. Pass through it, and enter smooth water.**
76	27.2	4½ min	I+	RIFFLE NO. 76	After some 300 yards of smooth water, a gravel bar extends into the river from the right bank, forming the riffle.
					HOW TO: Identify the 'V'-slot of this riffle (located approximately 20 feet from the left wall). Enter the 'V' and maintain position as the river curves around the wall, and the current carries the boat/raft to midstream. The current will carry your craft across the river to the right wall at the lower end of the riffle, and you may need to counteract the river's tendency to force you into the right bank.
---	27.9	---	---	*TATE CREEK CHECK- POINT*	*This is your twelfth CHECK-POINT. Tate Creek Check-Point is approximately 12 minutes drift time downstream from Riffle No. 76.*
					Tate Creek tumbles over a steep rock wall as it plunges about 45 feet to the river level below. Located on the north bank, the trail-bridge is also visible as it crosses the creek. This is one of the most tranquil spots on the river, and is a beautiful spot to have lunch. Approximately 200 yards up the creek there is an incredible natural slide which is so smooth you slide down it without abrasions, and you will drop approximately 25 feet into a deep, fresh-water pool. It is easy to imagine Indian children frolicking at this site many years ago.
---	28.0	---	---	SLIDE CREEK	Located about 150 yards downstream from Tate Creek, Slide Creek enters the Rogue on the left bank. A small sandy area adjacent to the creek makes a nice small campground or lunch spot. A high rock wall offers shade from the sun.
77	28.3	8 min	II+	TACOMA RAPIDS	Named after a mining corporation from Tacoma, Washington. Extensive mining was done in this area earlier in the century. A rockslide on the right bank has left an exposed, barren surface and many large boulders along the right bank. The resultant restriction in the river is the cause of Tacoma Rapids.
					HOW TO: Position yourself in the CENTER of the river. Pass down the center of the standing waves at the entry and the current will carry you to the left where you enter the main chute. Once into the main chute some 30 yards, MOVE LEFT to pass between two midstream boulders. After passing between midstream boulders, MOVE LEFT AGAIN to avoid a large standing wave directly ahead. Continue on in LEFT MIDSTREAM and adjust position to avoid obstacles on either side for another 100 yards.
---	---	---	---	CLAY HILL AREA	There are more bear in this area of the river than anywhere along your journey. They may be seen at anytime on either side of the riverbank. There are also a lot of otter in this stretch. Clay Hill is a privately-owned property which has kept the original homestead intact about 1 mile up the creek, above the river. The legendary Hathaway Jones' wife was born on this homestead. The remains of an old sawmill may still be found in the creek. Clay Hill Lodge is owned by Tom Staley, and is available for lodging and meals with reservations. Many of the jet boats from the coast stop here for food and rest.

What a ride!

Everything buried except the paddle.

Trouble at Blossom Bar.

"Orange Torpedoes" pick their way through Blossom Bar.

Hathaway Jones — he lived most of his life about 6 miles up the trail from Marial, on Bald Ridge. Hathaway married Flora Thomas, whose father homesteaded Clay Hill. Flora's younger sister is buried at Clay Hill. The homestead still stands today, about 200 yards up the trail from Clay Hill Lodge. Hathaway was the 'tall tale teller' of the Rogue River.

Clay Hill Lodge — Operated by Wild River Adventures. The lodge is open to the public. Reservations: (503) 826-WILD.

Clay Hill Rapids. Note how river sweeps left around gravel bar.

Rapid No.	Mileage	Drift Time	Class	Name	
78	28.6	7 min	III	CLAY HILL RAPIDS	***SCOUTING SHOULD BE FROM THE ISLAND IN THE CENTER OF THE RIVER***

Clay Hill Lodge is on the right bank, approximately 400 yards before the rapids. A boulder island splits the river into two channels — you want to TAKE THE RIGHT CHANNEL!

HOW TO: Enter the RIGHT CHANNEL. Position yourself approximately 10 feet off the LEFT SHORELINE OF THE RIGHT CHANNEL and maintain position as you are swept around the bend by the current. Avoid rocks and shallows in midstream.

| 79 | 28.8 | ½ min | II | LOWER CLAY HILL RIFFLE | This riffle is located approximately 80 yards below the end of the island at Clay Hill Rapids. |

HOW TO: Position yourself in dead CENTER and ride the crests of the standing waves approximately 100 yards. Then move slightly RIGHT OF CENTER to avoid large standing waves. At the end of the rapids, pull to RIGHT OF CENTER to avoid midstream boulder. Pass into quieter water.

| --- | --- | --- | --- | CLAY HILL STILL-WATERS | You now enter approximately 2 miles of stillwaters. Note the change in vegetation — southerly exposure has resulted in scrub oak and grasses growing in the dry environment. Rocks are composed of softer structure and their erosion is more dramatic, causing caves and caverns to be formed by the scouring actions of high water levels. |

Fall Creek Falls in the spring.   Courtesy Don Turcke

Flora Dell Creek and Falls — view from hiker's trail. Flora Dell Creek was named after Flora Dell Thomas, Hathaway Jones' wife.

Potholes in rocks along Clay Hill Stillwater. 'Drilling pebbles' located in bottom of potholes are swirled circularly by floodwater flows to create these curious holes.

Rapid No.	Drift Mileage	Time	Class	Name
---	30.2	---	---	*FALL CREEK FALLS CHECK-POINT*

This is your thirteenth Check-Point. Fall Creek Falls Check-Point is a varing drift time downstream from Lower Clay Hill Riffle. Depending upon whether the currents are swifter or slower, and whether you stop to sight-see some of the interesting features enroute, it may take from 10 to 30 minutes typical drift time.

Approximately 2/3 of the way through Clay Hill Stillwaters, you will find Fall Creek Waterfalls on the south bank. It tumbles almost 50 feet into the pool at its base. Another 100 yards downstream brings you to Flora Dell Creek on the north bank. It is well worth your while to pull ashore at both of these scenic attractions. At Flora Dell Creek, you can walk up to the hiker's trail where you will see the beautiful Flora Dell Waterfalls cascading over a sheer wall and entering a deep pool at its base. 200 yards below Flora Dell Creek on the south bank, an interesting live oak tree has exposed its root structures dramatically, as they search for sustenance along the face of a rock structure.

| 80 | 30.9 | 13 min | II | **PEYTON RIFFLE** |

Named after the original Peyton Ranch homestead. Sometimes Peyton Riffle is referred to as 'Slide Riffle' — but old-time river guides prefer the original name. A boulder island can be seen in the river. The MAIN CHANNEL FLOWS TO THE LEFT! Hicks Creek enters the Rogue from the north bank, approximately 200 yards downstream from Hicks Creek. Peyton Place Lodge is located on the south bank at the bottom of the rapids (it is also called Wild River Lodge), reservations (503) 826-WILD.

HOW TO: Position yourself in dead CENTER of the LEFT CHANNEL. Align with the center 'V'-slot. Work yourself a little (5 feet is enough) to the RIGHT OF CENTER. Maintain this position for 100 yards to avoid large standing waves (caused by submerged boulders). Pass midstream boulders closely (within 2 feet) on their right sides, and you will have no problems. At the lower end of the island, remain WITHIN 30 FEET OF THE RIGHT BANK to avoid rapid shoaling on the left side of the river. Pass the end of the island by riding crests of main flow to the end of the rapids.

Natural pavement stones found on south bank, about ½ mile below Peyton Riffle. 75

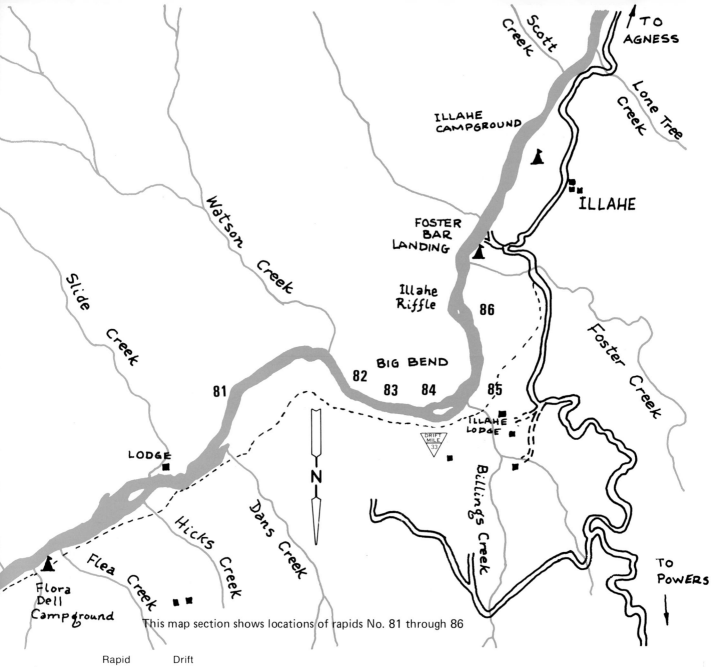

This map section shows locations of rapids No. 81 through 86

Rapid No.	Mileage	Drift Time	Class	Name	
81	32.0	10 min	II	**BURNT CREEK RIFFLE**	Named after Burnt Creek by old-time river guides, because Burnt Creek enters from the south bank at the end of the riffle. A large boulder bar extends into the Rogue from the right shore, forcing the river against the left bank, and forming the riffle. This is the beginning of Big Bend. **HOW TO: Enter the riffle in EXACT CENTER, then move across current to approximately 15 FEET OFF RIGHT SHORE and maintain this position downstream to the Watson Creek Stillwater stretch. As you pass the bottom end of the boulder bar, move to the LEFT side of the river to easily avoid obstructions on the right riverbed.**
---	33.1	---	---	**BILLINGS CREEK & ILLAHEE LODGE**	Billings Creek enters the Rogue from the right bank at the bottom of Brewery Hole Riffle. This creek was named after pioneer John Billings, who settled this area and lived several miles downstream at the mouth of the Illinois River. Illahee Lodge can be seen in the valley a few hundred yards from the river, on the right shore. The Lodge is privately owned and reservations are required. The canyon opens up into a wide valley in this section of the river trip.

Big Bend gets its name from the course that the river takes in this section of its travels. The Rogue is detoured from its southwesterly course as it swings through a large northwesterly semicircle. It returns to an almost due south direction of flow, and finally resumes its original southwesterly course towards the Pacific Ocean. This deflection is caused by the river encountering resistant rock layers, so it has changed its course to find the areas of least resistance. Big Bend is also the site of the last Indian battle of the Rogue River Wars of 1855-56. After the conflict at Battle Bar, General Lamerick's forces had driven the Upper Rogue Indians from their stronghold — the Indians were then driven downriver towards the Big Bend section of the river. The Indians were now trapped between General Lamerick's forces heading downriver, and Colonel Buchanan's entire company (343 troops) heading upriver. Colonel Buchanan invited the chiefs of the Indian tribes to a peace council held at Oak Flat (4 miles up the Illinois River). During these talks, held from May 19 to May 22, 1856, Chiefs George and Limpy agreed to surrender in a few days, and go to the reservation at Siletz. Chief John, however, refused by stating: "You are a great chief. So am I. This is my country; I was in it when those large trees were very small, not higher than my head. My heart is sick with fighting, but I want to live in my country. If the white people are willing, I will go back to Deer Creek and live among them as I used to. They can visit my camp, and I will visit theirs; but I will not lay down my arms and go with you on the reserve. I will fight. Good-by." Whereupon he took his departure unrestrained, as had been agreed upon.

Chief John pursuaded all the Indian chiefs to resist the white man's intrusion, and to fight for their territory. On May 27, 1856, the Battle of Big Bend began. It lasted for some 30 hours. This battle was the Army's most severe encounter in Oregon up to this time. Chief John's Indian warriors were winning the battle until the arrival of a reinforcement company of 54 soldiers. John's warriors found themselves surrounded and cut off from their canoes. The Indians escaped by crossing a steep ravine and climbing a mountain on the other side. Two days later, the Upper Rogue band surrendered to Colonel Buchanan at Big Bend. Chief John was the last to surrender. With only 35 warriors left, he was unable to fight for his land any longer.

Nearly 1200 Indians from southern Oregon were transported by steamer and land to the Siletz Reservation, 175 miles to the North. This ended the days of the Indian in the Rogue River Canyon.

Foster Bar

This section of the river is called Big Bend. It extends from Watson Creek Stillwater through Foster Bar.

77

Rapid No.	Mileage	Drift Time	Class	Name	
---	32.5	---	---	WATSON CREEK STILL- WATER	Named after Watson Creek, which enters a few hundred yards downstream. A low-water bridge crossed the river at this point. It was used by old-time loggers to remove timber from this area. On the north bank you can still see the remains of Buster Billings' old cabin.
82	32.8	12 min	II	**WATSON CREEK RIFFLE**	Named after Watson Creek, which enters on the left side at the head of the boulder bar which extends out from the left shore, causing the river to narrow to the right, and thereby forming the riffle. **HOW TO: Boulders and shallows restrict the river-flow to a 'V'-slot. Position yourself in the CENTER of the 'V'-slot and maneuver slightly to avoid obstructions on the left and right of the main flow.**
83	32.9	1 min	I	**RIFFLE NO. 83**	Located approximately 100 yards downstream from Watson Creek Riffle. **HOW TO: Position yourself in dead CENTER and ride through the unobstructed riffle.**
84	33.0	7 min	I	**BREWERY HOLE RIFFLE**	Named after the foamy 'suds' (developed by Riffle No. 83) which often accumulate in this section. An island can be identified in right midstream. Main river channel passes TO THE LEFT! River becomes VERY SHALLOW here. Billings Creek enters from right bank at bottom of this riffle. **HOW TO: Identify the boulder in the center of the river at the head of the left channel. Pass within 2 feet of this boulder ON ITS RIGHT SIDE to AVOID THE SHALLOW WATER ON BOTH LEFT AND RIGHT RIVER-BEDS. Remain near the RIGHT SHALLOWS POUR-OFF as the current carries you to midstream. Avoid boulders in riverbed. Pass out of the riffle in the center of the river.**
85	33.2	3½ min	I+	**BILLINGS RIFFLE**	A boulder bar extends into the river from the right bank. The river channel narrows to the left bank. **HOW TO: Position yourself 15 FEET FROM LEFT BANK. Maintain this position through the riffle. Pull off the left bank near the bottom of the riffle to counteract the current's tendency to push you onto a small ledge extending from the left bank into the river.**
86	33.4	3½ min	II	**ILLAHEE RIFFLE**	The word 'Illahee' is Chinook jargon meaning 'land on earth', which meant this was the land the Indians should fight for because they felt it was their own. Illahee Riffle is divided into two parts. The first part can be identified by the exposed, eroded high wall bank on the left side some 200 yards before the rapids, and by the island formation which can be seen in midstream. The main channel is ON THE RIGHT SIDE. Stay in the CENTER of the MAIN CHANNEL! **HOW TO: SECTION 1 — Remain in dead CENTER OF THE RIGHT CHANNEL until you reach some 100 yards downstream from the beginning turbulence of the island's neck-like shallows (WHICH CAN CUT YOU OFF FROM RE-ENTRY INTO THE MAIN CHANNEL!). At the lower end of the island, position yourself 10 FEET OFF THE LEFT BANK to avoid severe shallowing and pour-off on the right bank. As the left channel's water rejoins the mainstream's water, maintain your position near LEFT BANK along the crests of the standing waves. As you leave the end of the island, remain on the crests of the standing waves for some 300 yards until you approach Section 2 . . .**

Rapid No.	Drift Mileage	Time	Class	Name	
86 (continued)	33.4	3½ min	---	**ILLAHEE RIFFLE** (continued)	The river swings to the left of a boulder bar which extends into the river from the right.

HOW TO: SECTION 2 — Remain in dead CENTER of the river as you approach Foster Creek and boulder outcroppings on your right side. The current should carry you to the left bank as you pass by an old wooden bridge foundation structure. You may need to counteract the tendency of the river to force you onto the left bank. Once past this point, the river calms.

Rapid No.	Drift Mileage	Time	Class	Name	
---	33.8	5 min	---	FOSTER BAR & FOSTER CREEK	Your river journey is now ended. If you choose to continue downstream, there are several Type II rapids, but none are too difficult now that you have come this far, and you should be able to read them easily before drifting into them. They can be easily negotiated if you study them before proceeding. Several access points permit pulling out downstream. Foster Bar is the most popular take-out. A cement boat ramp is built here. You are now approximately 35 miles from Gold Beach, the confluence of the Rogue River with the Pacific Ocean. We hope you have had an eventful and memorable river journey, and that this River Log has helped make your trip more comforatble and secure.

Illahee campground is a large Forest Service campground located at Foster Bar. If you end your float trip late in the day, you may choose to camp here overnight. Old trenches and army fortifications are preserved in the campground. The Illahee/Big Bend area of the Rogue River was the site of the last major battles of the Rogue Indian Wars. Foster Bar and Foster Creek derive their names form Lt. Charles Foster. Foster escaped an Indian attack at this site and worked his way downriver to Port Orford (Fort Orford, in those days) to obtain help from troops located there. After the Indian Wars, Foster returned to this area, married an Indian woman, and had a daughter who in turn married another ex-soldier named Frye. The many Fryes who now live in and around the Rogue River canyon are probably descendants of Charles Foster.

This picture was taken in 1907. Two gillnet fishermen and their catch.

Alameda Mine.

Old mining equipment at
Whiskey Creek cabin.

Front half of breakaway boat
at Whiskey Creek cabin.

HISTORY AND NOSTALGIA

The first recorded visit of white men to the Rogue River was in 1825. At that time the Hudson Bay Company sent a group to explore the area and search for furs. The river was apparently given its name about that time by French trappers. They referred to the Indians in this area as 'Les Coquins' (The Rogues) and this river 'LaRiviere aux Coquins'(The River of the Rogues). The Indians of this entire area were hostile to the intrusion of the whites into their homeland. Jedediah Smith and his party of seventeen were attacked in 1828 on his first expedition North from the Smith River in California. Smith and three companions were the only ones to escape.

Southern Oregon was virtually unsettled wilderness until the California gold rush of 1849 occurred. Up to that time most of Oregon's population resided in the Willamette Valley. It is estimated that at the height of the gold fever from 2/3 to 3/4 of the male population of Oregon abandoned farms and family to head for California to strike it rich. Not all, however, were successful and prospectors began to look elsewhere for gold, including Southern Oregon.

In 1851 gold was found in Jacksonville and a new gold rush of considerable proportion followed. Then, in rapid succession, gold was found at Josephine Creek on the Illinois River and at Galice on the Rogue River. Josephine Rollins or Rawlins was the daughter of one of the miners and the only female in the area. Josephine Creek and subsequently Josephine County were named in her honor. Until 1856 Jackson County included what is now Josephine County. In 1855 it was the largest and wealthiest county in Oregon. But because of the difficulty encountered by people in the Illinois Valley area in traveling over the roads of that day to the county seat in Jacksonville, Josephine County was split off in 1856. In addition to the above mentioned gold strikes, many other very productive mines were discovered throughout the area. The Illinois Valley was particularly prolific. Althouse Creek on the upper Illinois was very rich as was Sailors Diggins. The latter was discovered by sailors who jumped ship at Crescent City, California to head for Jacksonville but struck it rich on the way. Other claims were staked out around Kerbyville (Kerby) and on the Applegate.

Galice was named for Louis Galice, a French doctor turned miner, who discovered gold deposits on that creek in 1852. Some activity has persisted in the Galice area intermittently to this day. Grave Creek was also mined extensively. Grave Creek was named for a girl, Martha Leland Crowley, who died and was buried there. It is thought that Indians later dug up her remains. It is also said that the Indians were later killed and thrown in the same grave.

Considerably later, 1880-1900, many other placer mines were operated along the lower Rogue canyon. Considerable activity was seen at such areas as Alameda, Rum Creek, Whiskey Creek, Tyee Bar, Black Bar, Battle Bar, Winkle Bar, Mule Creek, Blossom Bar and Solitude Bar.

Alameda mine, five miles below Galice, in addition to gold, produced silver, copper, lead and zinc. It was very active between 1908 and 1916. In 1940 it reopened, but closed in 1942 when the government closed down all nonessential mining activities by diverting labor and material to the mining of needed basic metals.

Gold was originally discovered by panning, but once established, placer mining took over. At first the rocker and sluice boxes were sufficient, but later hydraulic methods were utilized to tear up the ground down to bed rock. Flumes and ditches were constructed to convey water along hillsides to the mine. Most of these were constructed by Chinese Coolie labor. The water was brought in under 150-250 feet of pressure and passed through huge nozzles large enough to tear hillsides apart. Later, dredges were brought in to rework the areas. A part of one of these can be seen on the left bank at the head of Slim Pickens Rapids.

Later, as the placer deposits became worked out, lode deposits and ore mining became important. This required stamp mills or arrastra wheels to pulverize the ore. The arrastra was a large grinding device (round concrete disc eight feet or so in diameter) used to crush ore. One of these is located at the head of Solitude Bar in the underbrush about two hundred feet from the river.

Approximately $10,000,000 in gold was taken out of Southern Oregon between 1851 and 1860. In 1857 and 1858 as the placer deposits were beginning to decrease, a new gold rush occurred in the Fraser River area of Canada. Most of the miners followed the new rush and their places were taken by the Chinese who were willing to work harder and for smaller profits. It has been estimated that up to 3,000 Chinese miners were working along the Rogue River at one time, although this figure may be exaggerated. Total production of gold from Oregon in 1852-1960 was 134 million dollars, placing Oregon tenth among the gold-producing states, including Alaska.

Most of the heavy equipment for the mines in the canyon was brought in by way of the river, either downstream from Grants Pass, or upstream from Gold Beach. Barges were pushed, pulled, towed and motored as they floated the necessary implements to the gold mining sites. Many accidents happened, and much equipment was lost, as might be imagined. A five ton mortar box came downriver to Blossom Bar in a specially built riverboat. Parts of the two stamp mills at Blossom Bar came upriver from Gold Beach on a barge using a cable and hand winch.

The Indians in the area under consideration include the Takelma group on the upper river and from the Galice area, and the Tututni group of the coastal region. They resented the intrusion of the white men into their homeland and demonstrated their hostility early. The first expedition by Jedediah Smith north from California was attacked and only four of the seventeen escaped. They were resistant also to the Hudson Bay trappers which led to their name of 'Les Coquins' (The Rogues) and thus Rogue River.

Although the Rogue River Indian Wars are generally considered those of 1855-1856, there were many events of interest leading up to this final conflict. In 1851, when the gold rush started in Southern Oregon, there were many incidents between the miners and the Indians. Because of complaints that miners were being robbed, Joseph Lane (Governor of the Oregon Territory and superintendent of Indian Affairs) came to the area and was able to make a treaty with the Indians. Lane was apparently an honorable man with a strong personality and the Indians respected him. The chief, in fact, requested to be allowed to use Lane's name and was thereafter known as Chief Jo.

The Indians followed the treaty much better than the whites, but continued ill treatment and wanton killings ultimately led them to retaliate in kind. Part of the reason the whites felt free to usurp the Indians' homeland was that Congress in 1850 passed a law extinguishing Indian titles to lands West of the Cascades, in order to award clear titles to settlers moving into the area.

The winter of 1852-53 was reasonably peaceful, due to the severity of the weather. All roads into Southern Oregon were blocked and what supplies there were became precious. Flour cost one dollar per pound, tobacco fetched one dollar per ounce and salt traded ounce for ounce with gold.

Difficulties were experienced by local settlers in obtaining regular army support, so they joined with the miners and formed volunteer groups and elected their own officers. These volunteers not infrequently perpetrated atrocities equal to any the Indians were guilty of. On one occasion near Yreka, Indians were invited to a feast and the meat was poisoned with strychnine. On another occasion, seven Indians were invited to dinner at Grave Creek to celebrate a peace settlement, and once inside the house they were shot down in cold blood.

Early in 1853 seven miners disappeared near Galice and, although no bodies were found, Indians were considered guilty and those suspected were tried and hung. The Indians retaliated in kind and many killings on both sides occurred during the summer.

In August of 1853, two battalions of regulars and volunteers located and surrounded the Indians on Evans Creek. Joseph Lane was in command and in the fight that followed was wounded in the shoulder. Although far from beaten, when the Indians learned that 'Jo Lane' was present, they agreed to stop fighting and attend a peace council. On September 10, 1853 a treaty was agreed to at upper Table Rock near what is now Bybee Bridge close to White City.

By the terms of the treaty, the United States acquired the whole of the Rogue River Valley. The Indians were given a reservation around Table Rock and forty-five thousand dollars, to be given periodically in goods and implements. According to the secretary of war, the Indian disturbances in Southern Oregon in 1853 cost the lives of over one hundred white persons and several hundred Indians. This estimate included the Yreka area of Northern California.

In 1854 further incidents occurred but not as many as in previous years. It is of interest that the official reports of the Indian Agents of the United States Government tended to indicate that the Indians were suffering grievous wrongs and that 'in no single instance have the Indians been the first aggressors.' It is also of interest that the U.S. Army generally treated the Indians more fairly than did the local citizens' groups and the volunteers. General Wool, in command of the division of the Pacific, in sending a reinforcement to Fort Lane, stated that it was to protect the Indians against the white man and not to protect the settlers. One spokesman for the volunteers (Mrs. Victor) has stated that the volunteers believed in preventing robbery and murder rather than chastisement after the crime was committed. Presumably, prevention was accomplished by annihilation of all Indians.

The Rogue River Indian War of 1855-1856 was actually launched on October 8,1855. Before daybreak of that day a volunteer group organized in a Jacksonville saloon attacked without warning an encampment of peaceful Indians at the mouth of Little Butte Creek. Thirty Indians were killed, eighteen of them women and children, and the rest old men.

In retaliation the following day, October 9,1855, Indian bands pillaged up and down the valley from Evans Creek to Grave Creek, killing upwards of twenty persons, again including women and children.

The war was on. Many volunteer companies were formed immediately in the thoroughly alarmed villages and mining camps. The first engagement between the volunteers and Indians occurred at Skull Bar, just below the mouth of Galice Creek. The buildings in Galice were burned and nearly one-third of the company were killed or wounded. The settlers had been warned of the impending attack by Umpqua Joe, a friendly Indian. In recognition of this, in 1885, the United States Government gave a parcel of ground to Joe's daughter Indian Mary. This was known as the smallest Indian reservation ever created, and today serves as Indian Mary Park and Campground, a mile below Hellgate Bridge.

Two weeks later scouts located the Indian encampment in the hills at the headwaters of Grave Creek, near the Cow Creek divide, approximately four miles northeast of the present Grave Creek bridge. A force of four hundred volunteers and one hundred and five regulars attacked the stronghold on October 30 and 31st and there was fought the Battle of 'Hungry Hill' and 'Bloody Spring.' In two days of hard fighting the white forces suffered a sharp defeat and lost approximately forty killed and wounded.

Following this battle, the Indians withdrew downriver. Here their camps were on various river bars where they were protected on one side by the river and on the other by steep wooded hills. On November 26,1855, a force of four hundred thirty-six volunteers and regulars tried to attack a camp of two hundred Indians at Black Bar, but were unable to cross the river against the Indian rifle fire. Bad weather prevented reinforcements from arriving and after several hours the whites withdrew with a loss of one dead and four wounded. The rest of the winter was relatively peaceful. The Indians moved their camps to the 'Big Meadows' area which lies between Horseshoe Bend and Mule Creek. Some scattered Indian camps on the Applegate and on Little Butte Creek were attacked and destroyed during the winter.

After the massacre on October 8 at Little Butte Creek, the Rogue River Indians were able to convince the coastal Indians around Gold Beach and the Cow Creek Indians, that the settlers meant to exterminate all Indians in the area. These groups then joined in the hostilities. After the winter the war can be considered in two parts — the canyon area and the coastal area. The general plan was to have armies coming downriver and upriver to trap the Indians in between.

In the Gold Beach area the Indians struck a surprise blow on February 23, 1856. They lured Indian Agent Ben Wright and Captain Pollard to their camp where they were murdered. The Indians then attacked the volunteers' guard camp and killed eight of the ten men there. The people in and around Gold Beach, one hundred and thirty in all, moved immediately to an unfinished fortification 'Miners Fort', about one and a half miles north of the mouth of the Rogue River. The Indians then looted and burned every house on the river, sixty in all. Twenty-six persons were killed. Later six more were killed when soldiers from the fort tried to get to a potato cache along the riverbank. The siege of Miners Fort lasted thirty-one days until relieved by two companies of United States troops marching up from Crescent City.

In mid March a modest skirmish occurred near the mouth of the Pistol River involving a group of thirty-four volunteers. Other encounters also took place in the Illinois Valley, Cow Creek and along the Coquille River.

The next major engagement was a result of a plan to attack the Indians in force at their main camp around Big Meadows. A total army of regulars and volunteers of five hundred and forty-five men finally located the main camp at Battle Bar. The troops were able to approach under cover of a fog bank, but were unable to cross the river. After considerable shooting back and forth the Indians abandoned their position. A fort was then constructed at Big Meadows called Fort Lamerich. This was in keeping with the total war plan to drive the warring tribes down river away from the valley settlements while forces under Colonel Buchanan worked upriver from the coast.

A significant skirmish that shortened hostilities occurred at Shookum House Mountain above Lobster Creek, when Indians were driven from an ill-conceived fortress with a loss of sixteen known dead warriors. Another step in the decimation of the enemy occurred at the mouth of Lobster Creek when a detachment of volunteers spotted two canoes filled with Indians coming down the river. They quickly hid behind a large rock and waited. As the canoes passed, the soldiers opened fire and in a few minutes killed twelve Indians including one woman. The rock is now aptly called Massacre Rock.

In May, after the Battle of Big Meadows, Colonel Buchanan moved his force of regulars to Oak Flat near the mouth of the Illinois River. The Indians were invited to attend a peace council which was held on May 21, 1856. All the chiefs except John seemed ready to give up.

"You are a great chief," said John to Colonel Buchanan. "So am I. This is my country. I was in it when those large trees were very small, not higher than my head. My heart is sick with fighting, but I want to live in my country. If the white people are willing I will go back to Deer Creek and live among them like I used to. They can visit my camp and I will visit theirs; but I will not lay down my arms and go with you on the reservation. I will fight! Good bye."

The other chiefs agreed to surrender on May 26th, and Captain Smith was to meet them with his eighty dragoons. The Indians, however, came not to surrender but to fight. Captain Smith had been forewarned to expect an attack and had moved his troops to high ground. Even so, after two days of fighting, his troops were in desperate condition when finally relieved by reinforcements. After this battle, which took place at Big Bend near Illahee on the north bank of the river, all the chiefs except John surrendered. John's tiny band of thirty-five or so warriors, however, was no match for one thousand soldiers and he finally surrendered on June 29, 1856.

Of six hundred Rogue River and Cow Creek Indians removed to the Siletz reservation, approximately one-half died within one year.

Indian Agent Ben Wright - killed in Gold Beach massacre.

Chief John.

Col. Robert C. Buchanan.

Andrew Jackson Smith - Commandant at Fort Lane.

"The Man and the River"

Glen Wooldridge made his first float trip down the Rogue River in 1915. At that time a man was considered to be insane to venture down the impossible river. It was not uncommon to take a full day to portage around Blossom Bar. Since the early days Glen has done much to make your trip safe and more enjoyable. Most of the major rapids have been blown with dynamite to make the river passible. In 1947 Glen was the first man ever to motor up the river from Gold Beach to Grants Pass.

DUTCH HENRY

There used to be a man who lived on the river at Horseshoe Bend. Only the real old timers remember him. He was known as Dutch Henry to the river people. He was an intermittent miner but most folks recalled him as a friendly old man who raised cattle and did some packing. He established his own trail called the Dutch Henry Trail from Little Meadows to Glendale where he got supplies for his little trading post. He was most widely known for the cattle he owned that he let run wild in the hills. The river residents seemed to suspect his past was somewhat clouded, but few knew the real story.

According to the recollections of an old time sheriff of Curry County two men, Scotty McMullen and Henry Rosen or Rosenbrook, had a small trading post about 60 miles up the Rogue River from its mouth. Scotty got married and wanted out of the partnership. They argued about the division of property and Scotty wound up dead.

Henry later claimed that Scotty came after him with a hatchet and that he had hit Scotty in the head with his rifle and killed him. Unknown to Henry, a couple of neighboring miners had heard the row and slipped up close enough to witness the fight. They were to be the prosecution's witnesses.

During the inquest, the body was dug up for examination and it was noted that Scotty's head had been beaten to a pulp. The trial was held in Coos County in 1876 but the witnesses never showed up. The jury had to accept Henry's story and he was freed.

Henry returned to his trading post on the river. His next altercation was with a man named Black for whom Black Bar is said to be named. There seems to be some question of the exact sequence of events. Some sources say that Black ran off with Henry's Indian wife. Others indicate Black was a miner who tried to get Henry to pay back some money he owed him. In any event, a body later washed up on the beach which was identified as that of Black. Black's head was found to have a hole in it made by a knife blade and evidence suggested that Henry was the likely suspect.

He was brought to Gold Beach for trial in 1881. The prosecution did not have much good evidence and the sheriff was requested by the prosecuting attorney to bring Black's head to the court room for evidence. Henry told the court that Black was drunk when he left his store and that he later saw the man's skiff floating upside down on the river and that Black must have fallen into the river and drowned. The prosecutor wanted to show that Henry's knife blade fitted the hole in the skull, but when the sheriff came into court with Black's very smelly head in a box, the judge declared that he would not allow any such evidence in his court. So Henry was free again.

Henry was later under suspicion for two other mysterious disappearances. One was a man who told friends he was going to stay overnight at Henry's and was never seen again. Another was found dead in his burned cabin after Henry had spent the night with him. Evidence, however, was slim and nothing was ever done about either case.

Dutch Henry lived out his days at Little Meadows below Horseshoe Bend and still remains the mystery man of the Rogue River canyon.

JIMMY COE

Jimmy Coe was a strange little man who was familiar to river residents from Black Bar to Marial. He arrived in the area in 1927, living most of the time in a shack on Missouri Bar. River guide Sid Pyle said that he had a summer home and a winter home. The summer home was a piece of tarp and boards near the river. In the winter he lived in a cabin a short distance up the creek. He was a small man, 5 feet tall, and weighed about 90 pounds, and was very cantankerous. He always carried a gun which acquaintances said was too old and rusty to shoot. People drifting on the river often became nervous when they would see him watching them from behind rocks and trees. He had quite a reputation for stealing things that were not nailed down. Sid Pyle tells of the time he stole some planks from the Anderson Ranch. Anderson left a note for him not to take any more. Coe wrote under that note: "Go to hell you S.O.B."- signed, Jimmy Coe.

In 1953 he threatened a man at the Post Office in Marial and pulled his gun which failed to discharge. He was disarmed and fled into the woods. When captured, it was learned that he was an escapee from Florida State Prison where he had been serving a sentence for murder. His real name was James Johnson, and he was returned to the State of Florida.

Tate Creek Slide
Courtesy Sundance Expeditions

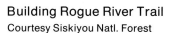
Building Rogue River Trail
Courtesy Siskiyou Natl. Forest

"AMBUSHED"

Clarence Burke created some excitement on the river in 1934. Burke was a miner who lived in a cabin at the mouth of Wild Cat Creek on the south side of the river. Below him a ways at Howard Creek lived two young miners, Roland Burr and Martin Jennings. In those days there was a trail on the south side of the river that the miners used but which has since mostly been erased by floods.

Burke apparently had reason to suspect that Burr and Jennings had been stealing supplies from his cabin. In April of 1934 Burr and Jennings trespassed onto Burke's property, not knowing he was at home. As they started up the steps to his cabin, Burke shot them both. Burr was wounded fatally but Jennings escaped and made his way to the Rand C.C.C. camp where he reported that he and his partner were ambushed as they were trying to get a drink of water.

A posse was formed but when they arrived at the scene of the shooting Burke was gone, having walked to West Fork and caught a train to California. Burr's body was recovered from the river where Burke had thrown it after first wrapping it in canvas and weighting it with rocks.

Burke was later apprehended in California and returned to Grants Pass for trial and found not guilty.

THE TALL TALE TELLER OF THE ROGUE

Surely one of the most remarkable characters that ever lived and labored in the Rogue Canyon was Hathaway Jones whose reputation as a teller of tall tales far exceeded the limited area and audience to which he had access. Glen Wooldridge verifies his remarkable talents.

Hathaway and his father, William Sampson "Samps" Jones, came to try their hand at mining in 1890. They worked a claim on the East Fork of Mule Creek for two years and then moved to Battle Bar where they lived until the 1920's when Sampson had to leave because of old age and illness.

Sampson and Hathaway Jones place at Battle Bar.

Courtesy Siskiyou Natl. Forest

In those days much of the supplies and mail was brought into the Rogue Canyon by pack train from the nearest railroad station which was at West Fork in the Cow Creek Canyon. The trail came 21 miles from West Fork to Marial by passing through Big Meadows which was the post office before it was moved to Marial in 1903. From Marial the trail went on down river to Illahe and Agness.

In 1898 Hathaway was given a contract to carry mail and supplies from West Fork to Illahe and he continued this lonely work for nearly forty years. He married Flora Thomas whose parents were homesteading at Clay Hill and for whom Flora Dell Creek and Falls is named. He took up a homestead along the trail above Big Meadows near Bald Ridge.

Hathaway Jones pack train at Marial

Courtesy Siskiyou Natl. Forest

The fantastic stories with which he entertained the river people were presumably dreamed up and embellished during the long hours on the trail. A collection of his stories is available in a book entitled "Tall Tales From Rogue River" by Steven Dow Beckham.

In 1937 he was killed along the trail above Illahe. The saddle on his horse came loose and he fell off and broke his neck. He was buried in the cemetery at Foster Bar.

THE INCREDIBLE RESCUE

The account of the rescue of a downed pilot by Glen Wooldridge has to be of interest to any river drifter.

On December 27, 1955, Fred Hale, veteran Grants Pass pilot, was making an air drop of meat and supplies to Black Bar Lodge when a gust of wind tipped his light plane into the mountain side. The Lodge caretaker, Red Keller, took him to the Lodge badly burned but he had no way to communicate to the outside world.

This was all during the famous 1955 storm and flood. The river was 20 feet above normal level. Search planes could not see through the dense clouds and helicopters were grounded. Ground teams tried to bulldoze their way to Marial but could not get through the deep snow.

Glen and his son Bruce put their boat in the water at Alameda to search the river banks. The boat was powered by a 20 horsepower motor and they reached Black Bar in 70 minutes, having passed right over Rainie Falls on the way. Finding Hale at the Lodge, they rigged a cot in the bottom of the boat for the wounded man and started for Marial in the pouring rain and on what Glen described as "the roughest water I ever saw in my life."

The most severe turbulence was encountered at Horseshoe Bend where the motor drowned and the boat nearly swamped. The jolting was so great that the cot collapsed, dropping Hale into 8 inches of water in the bottom of the boat. Fortunately, they were able to get the motor started again and made it to Marial where Hale was cared for at Marial Lodge until he could be taken out by helicopter three days later.

Old pack bridge above Rainie Falls.
Courtesy Siskiyou Natl. Forest

THE SECLUDED BRIDGE

There used to be an old swinging foot bridge spanning the Rogue River just above Mule Creek. From there a trail went up the mountain to Bear Camp. This bridge was built by the Forest Service before 1920 and removed in the fifties because it was unsafe. George Morey tells how the bridge figured in the capture of a bank robber in 1924.

The bank in Glendale was held up by a lone bandit who then headed for the coast by trail. The authorities surmised that he would cross the river on this foot bridge to avoid the main traveled trails. Two deputies from Glendale went to West Fork by train, then hiked to Marial by trail. They then concealed themselves at the end of the bridge and on the second day made the arrest, recovered the money and hiked back to Glendale with their prisoner.

Old Suspension Bridge at Marial
Courtesy Siskiyou Natl. Forest

THE HIGH PRICE OF MEDICINE

George Morey, now retired in Gold Beach, used to be a ranger at Marial and also at Agness. He has been a source of some interesting stories of local characters. One such was Slim Damon who was a miner on Mule Creek.

Slim carried his .38 caliber pistol in a shoulder holster and one day, while bending over, his gun fell out and discharged. The bullet entered his body in front and came out near the backbone. Slim at first thought someone had shot him in the back.

Being alone at the scene, he fired three rapid shots as a signal and then went into his cabin and laid down. Another miner, Charles Lewis, heard the shots and came to investigate. He sent word to Marial where a telephone was available and West Fork was notified that a doctor was needed.

Dr. Faucett of Glendale drove to Bald Ridge in his car and then was taken by mule seven miles to Slim's cabin where he did what he could for the victim. During the night Slim aroused enough to inquire what the doctor's fee was and was told that it was $25.00. Slim let it be known that he thought this was outrageous but, inasmuch as he was still in critical condition, he paid up. The next morning he was carried by stretcher the seven miles to the road, and then by ambulance to the Roseburg V.A. Hospital where he eventually recovered.

Strangely enough, Slim passed away nine years later at the exact place of the accident, but this time of a heart attack.

Old Blossom Bar before blasting.
Courtesy Milo's Sporting Goods

Red River Mining Co. Flume at Mule Creek.
Courtesy Siskiyou Natl. Forest

Alameda Mine & Smelter & Low Water Bridge
Courtesy Siskiyou Natl. Forest

MAMMALS OF THE ROGUE

As you adventure down the Rogue River you will have the thrill of seeing many interesting animals along the river's edge. The following animals may be seen as they prowl along the side glens and ridges of the Rogue River Canyon:

BLACK BEAR

The black bear is the only bear known to exist in Oregon today and, although some scientists classify two separate subspecies, they are very similar. At one time the grizzly bear inhabited the Rogue River canyon, but no longer exists in the state. Confusion in identification of the black bear often arises because of the various color phases. A female may have a litter of three cubs, each with a different coloration. Colors vary from buffy-tan to cinnamon to black. One main characteristic that distinguishes the black bear from his larger relative, the grizzly, is his claw structure. The black bear has short, curved claws which make it fairly easy for him to climb trees. On the other hand, the grizzly has longer, less curved claws, and is not as adept at taking to the tree-tops. Also, the black bear has a 'Roman' nose, and does not have a prominent hump on his shoulders.

The total length of the black bear is about 4 to 6½ feet, and the average weight is from 200 to 300 pounds, with an exceptional one weighing up to 500 pounds.

The black bear is an omnivorous feeder — that is, he feeds on both plant and animal life. He will eat almost anything available, alive or dead. Insects, live or dead fish and game, berries, and many kinds of plants, including leaves, stalks and roots all make up part of his diet.

Normally, he is an independent animal and minds his own business, but occasionally gets into trouble when he mixes with civilization. Often in Oregon he does not sleep for the entire winter and will roam about seeking food during the warmer spells of weather. Cubs are born about January and weigh only 6 to 8 ounces. The cubs are blind, hairless and toothless at birth.

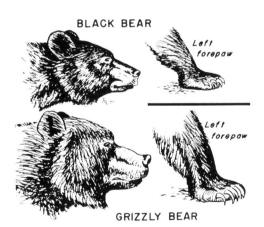

BLACK BEAR

Left forepaw

Left forepaw

GRIZZLY BEAR

The best places to see bear along the river are between Horseshoe Bend and Clay Hill. During the summer months black bear prowl along the bank looking for dead salmon, lamprey and whatever else they may find. In early fall, the bears will leave the river and return to the mountain slopes in search of ripe berries.

COYOTE

Whereas his larger relative, the wolf, has failed to adapt to civilization, the coyote has succeeded remarkably well, and at present Oregon still has a sizable coyote population. He is found in various color phases, the eastern Oregon one being generally lighter. The darker western Oregon coyote occasionally is mistaken for a wolf; however, the smaller size is a reliable identifying feature.

The coyote averages about 30 pounds, or one-third the weight of the wolf. Total length is about 4 feet, and height at the shoulders is 12 to 18 inches. He is the songster of the canyon and you may awaken some evening to his lonesome tones.

Because of his wide distribution and abundance, the coyote is the most destructive of the mammals of prey in the state. Though he takes domestic stock, particularly sheep, he also serves as a control on rabbit and rodent population; consequently, these smaller animals may become pests when coyotes are too severely hunted.

Coyote pups are born in April or May and the average litter is about 7. The coyote has been known to cross with domestic dogs and often the offspring is a large, vicious, animal. Some of these have been identified, mistakenly, as wolves.

As you journey down the river you may see coyotes ranging along either bank in the morning hours or late in the evening as they come to drink from the waters.

COUGAR

This largest member of the cat family in the state is found in most timbered areas where civilization has not forced him out. Variously called "mountain lion" and "panther", this big fellow is shy by nature and seldom lets his presence be known unless he is actively pursued. When cornered, he is a match for the best of dogs. Man is about his only serious enemy.

COUGAR

The cougar in Oregon varies in color from yellowish brown to dark reddish brown, depending on the area in which he lives. The western Oregon resident is generally darker. The distinguishing feature is the long body and tail, the total of the two often reaching 8½ feet or more. He is the only large, long-tailed cat in Oregon.

Despite reports to the contrary, a cougar seldom attacks humans and instead will make every effort to avoid meeting one. He becomes a nuisance when he develops a taste for domestic cattle and horses, but more often he avoids civilization.

Young are born at most any time during the year and an average litter numbers two.

Since he does not hibernate, the cougar must hunt for food all year. Food habit studies show that deer are the animals most often killed and eaten. Rabbits and porcupines are also important food for him. Other animals that he kills are raccoons, skunks, antelope and elk. An old cougar will kill domestic animals.

In over 100 trips down the river we have only seen one mountain lion. The big cat watched us from the rock bluffs on the south side of Blossom Bar as we dodged the boulders in our drift-boat.

BOBCAT

The other most common member of the wildcat family in Oregon is the bobcat. He gets his name from the short bobtail, the only other member of this family having a similar tail being the lynx.

Bobcats are found widely distributed in Oregon from the coastal mountains to the dry rimrock areas of southeastern Oregon. Those found east of the Cascades tend to be lighter colored than the western type. The bobcat is yellowish brown in color, with dark spots on most of his body. Total length is about 3 feet and height at the shoulders about 14 to 16 inches. His weight varies from 20 to 30 pounds. The only other cat that might be confused with the bobcat is the lynx. The bobcat has considerably smaller feet and shorter ear tufts than the lynx, and the lower surface of his tail has a white tip, whereas the lynx has a complete black tip on his tail. Also, the over-all size of the bobcat is smaller.

The bobcat ranks with the coyote as a predator of small birds and mammals, and occasionally sheep and other domestic livestock. His diet includes goodly numbers of rodents and he wages war on domestic house-cats that have gone wild. Generally, normal trapping pressures keep bobcat numbers under control, but in some areas trappers are employed and bounties are paid by some counties as an inducement to hunters.

Normally, the bobcat hunts by night although he is seen quite often in the daylight hours. He seldom backs off from any other animal, and presents an unfriendly attitude. The average litter is born in March or April, and usually numbers three.

BLACK-TAILED DEER

Easily recognized by its black or dark-colored tail with white undersurface and edges, this Pacific Coast deer prefers the brushy, logged-over lands and the Douglas Fir forests of western Oregon. During the fall, the predominant body-color is brownish-gray. The summer coat is reddish-brown.

Antlers of the buck usually branch in pairs and have brow-tines of medium length. These are shed in January or February and new growth starts a couple of months later. Antlers are covered with a soft, velvety skin until growth is completed in late summer and the buck polishes them clean in preparation for the mating season which comes in November.

Fawns are generally born in June and twins are common. They are reddish-brown with white spots on their back and sides. This protective coloration remains until August or September when the young ones take on adult pelage.

Food habits of black-tailed deer vary somewhat between the northern and southern parts of their range in Oregon. They are browsing animals feeding on various woody plants and small trees. In the Coast Range, trailing blackberry, red huckleberry, elderberry, vine maple, willow, and similar plants form a high percentage of their diet. Grasses are eaten in spring and early summer and fireweed and other weedy annuals are also consumed. Like other deer, the blacktail is choosy about his food, preferring to nibble a little of first one thing and then another, keeping an ever-watchful eye open for intruders.

As openings become overgrown and closed-in by taller trees along the riverbanks, the deer are forced to seek more open areas. As you journey down the river you can expect to see blacktail deer along the sandy beaches and gravelly bars. During the heat of the day, they will often be seen bedded-down in the shade of large trees or rocks.

ROOSEVELT ELK

This animal, sometimes called the Olympic elk, is an inhabitant of the rain forests of the Pacific Coast. In Oregon, it is most commonly found along the coastal range of mountains, but also occurs in portions of the Cascades. It is larger and heavier than the Rocky Mountain elk and antlers of the bulls tend to rise straighter from the head and be less widespread.

Almost no seasonal migration takes place in herds of Roosevelt elk. Even serious food shortage fails to move animals to new pastures. Hunting pressure has undoubtedly been responsible for some spreading, but many herds are still found where they were first discovered. This tendency of the herds to remain localized has concentrated hunting in certain areas, making it necessary to protect the animal through regulation.

Openings in the forest where logging or fire has encouraged growth of shrubs, grasses and weeds have become favorite haunts for Roosevelt elk. They eat many of the same browse plants as deer but also feed extensively on sedges, grasses and weeds. Fir thickets become hiding places when such shelter is needed, and despite their large

90

size, elk can move through such cover with surprising speed and silence.

September is the main breeding month and mature bulls gather harems of cows which they jealously guard against intruders.

During the summer months along the river, the elk remain at higher elevations. As you make your return road-trip from Foster Bar back to Galice, you will probably see them as you travel along road number 3400.

Most of the elk are found below Blossom Bar where the canyon opens and provides good grazing areas for these huge animals.

OTTER

One of the clowns of the animal world is the otter. For indicating sheer enjoyment of living and delight in being alive, this frivolous animal takes the prize. Some of his activities include: follow the leader, tag and sliding down mud or snow-covered slopes into the water. His impishness makes him a disgusting pest to the hard-working beaver trying to build a dam.

Otter are found on a majority of the streams in Oregon but are not numerous in any particular area. Whenever possible, they do their traveling by water and are very well adapted to this system. Their short legs and webbed feet make them excellent swimmers but rather poor 'landlubbers.' When crossing land they travel with a long loping

gait and, if forced to, can outrun a man for a short distance.

The fur of the otter is one of the best natural furs available, because of its dark color and density of pelage. It is equalled only by the fur of its cousin, the sea otter.

The general appearance of the otter is unlike any other animal, except perhaps the weasel. However, the otter is 3½ to 4½ feet in length, and measures 9 to 10 inches at the shoulders. Its weight averages from 10 to 25 pounds, with the females being generally smaller.

Coloration of both sexes is a rich dark brown with underparts, throat and muzzle often grayish.

Otter feed on shellfish, fish and other animals, but at the present population numbers, they are seldom a real nuisance except in certain specific instances.

The normal family size is two or three otter pups, born in April in a bank burrow or in an abandoned beaver or muskrat den.

Otter may be seen along the shore or in the back-eddies from Grave Creek to Foster Bar. They are most abundant between Blossom Bar and Big Bend. Keep a sharp lookout for them, they are easy to miss.

RACCOON

The masked prowler of the night is an apt name for the raccoon. One of the most distinguished characteristics of this friendly rover is his black mask. This, coupled with a ringed tail, makes it quite doubtful that he can be confused with any of our other fur-bearers.

In feeding habits, the raccoon is not unlike human beings, since he will eat almost any type of food. One of his most well-known habits is the one of washing or dunking food before he eats it.

When food gets too scarce during the winter, he will often curl up for long periods of sleep, though this is not true hibernation as with ground squirrels and marmots. At times, several families will utilize the same den.

The raccoon is mainly a nocturnal wanderer, but can often be seen during the daylight hours along stream banks. Economically, he is generally more beneficial than harmful, though his lack of fear of civilization often makes him a pest. Night raids on chicken houses and forays into corn patches are not uncommon by individual animals that acquire a taste for delicacies thus obtained.

Raccoon are found scattered throughout the state, and the eastern Oregon subspecies is the largest variety in the United States.

The average raccoon is from 10 to 16 inches tall at the shoulders. An average weight is 10 to 25 pounds, with occasional large individuals reaching close to 50 pounds.

The over-all color of the typical raccoon is grayish-black, with the black mask and ringed tail being most prominent identifying markings.

Young raccoon are born in April or May, with the average litter being four or five. Raccoon tracks may be seen on almost any sand bar along the river. Their nocturnal habits will bring them into your camp at night, so make sure you cover your food to avoid morning surprises!

GRAY DIGGERS

Another name for this interesting animal is the 'Douglas Ground Squirrel.' This squirrel resembles the silver gray squirrel in size and general appearance, but tends to have more brownish coloration mixed with its gray. Ground squirrels are also more sparsely furred and do not have the thick bushy tail of their tree-dwelling relative. Gray diggers will climb trees in search of food, but rush to their ground burrow if alarmed.

Burrows are often dug under the edge of downed logs, stumps or rocks, but will sometimes be found in open fields or meadows. The young, numbering five to eight, usually appear in the latter part of June.

Because of their large size and habit of feeding on roots, bulbs and other green vegetation, 'diggers' can become a nuisance in local areas, but are easily controlled.

As do several other members of the rodent family, gray diggers put on a thick layer of fat during the summer and go into hibernation for four or five months during the coldest weather in the winter. Warm days in western Oregon will bring out a few light sleepers, but generally the snooze is uninterrupted until spring.

These little guys are found all along the river. They will come into your camp and sample your food if invited. The fattest and sassiest one on the river is found at the small campsite located at Sports Illustrated Rock, above Battle Bar.

RING-TAILED CAT

The least conspicuous, but most nicknamed cat along the river is the Ring-Tailed Cat. It has also acquired the names of 'cacomist', 'ringtail', 'miner's cat' (in the olden days the miners used to tame this animal as a household pet), 'Bassaris' and 'little fox'.

With his large inquiring eyes and delicate, tapered face, the ringtail is one of the most appealing of the furbearers. His temperament is more gentle than that of his musk-bearing relatives.

The diet of the Ring-tailed Cat is much like that of his cousin, the raccoon, consisting of a large variety of plant and animal material.

Like the raccoon, the ringtail has a series of black bands around the tail. However, unlike his cousin, the bands on the ringtail do not meet on the underside as they do on the raccoon.

The ring-tail is much smaller than the raccoon and does not have a black mask across the face. He is a slender animal from 25 to 30 inches long and about 6 inches high at the shoulder. General over-all color is brownish yellow with whitish shades around the cheeks and lips. Weight is about 2 or 3 pounds.

The young are born in May or early June, with an average litter size being three to four kittens. The nest is usually built in a rocky crevice or hollow tree.

The only ringtail we have seen in our adventures down the Rogue was at Doe Creek on the high bank above the campground on the south side of the river.

OTHER SMALL MAMMALS

Others include the procupine, chickaree, ground squirrels, flying squirrels, chipmunks, skunks, weasels, badgers, muskrat and beaver.

Osprey

BIRDS OF THE ROGUE

BLUE HERON

The stately blue heron can be seen along the banks of the river. This bird is often mistakenly called a crane, but can be distinguished by its bluish-gray plumage, white head with black crest, and dagger-like bill. Individual

specimens vary considerable in size, ranging from 40 to 48 inches in length and having a wingspread up to 72 inches. When in flight, the heron always flies with a recurved neck. Although the principal food of the blue heron is various kinds of fish, its diet also includes snakes, frogs, mice, salamanders and insects. It fishes by night as well as by day, employing two different methods: still-hunting and stalking.

KILLDEER

The killdeer commonly breeds along the gravel river bars of the Rogue. It may remain through the winter

unless snow or ice forces a retreat to warmer climates. A medium sized plover, it is readily distinguished from other shorebirds by the two black bands across a snowy breast and the rufous rump and tail patch. The name 'killdeer' is derived from its distinguished alarm call, "kill-dee" which it utters at first sign of danger and continues until the intruder has past.

OSPREY

The osprey is found throughout the state, though not in great numbers. In some areas of the eastern United States this bird is strongly protected and even provided with artificial nesting sites. He feeds mainly on fish, and such species as carp and suckers make up a large part of his diet. Wings are long and pointed and the bird has an over-all white appearance underneath, with a dark, slate blue back. The tail is tipped with white and has six or seven narrow blackish bands. Wingspread is 54 to 72 inches and body length is 21 to 25 inches. As would be expected, the feet are large and the toes arranged with one reversible for better grasping. He is usually found near water and often can be seen dropping down to the water taking fish. Rarely does he eat other types of food such as rodents, frogs and birds. He is likely to be seen anywhere along the Rogue.

Red-tailed Hawk

RED-TAILED HAWK

Few hawks take the abuse that this one does. Usually called 'chicken-hawk', he is the most commonly seen and the most often killed. He frequently takes rodents near barnyards and hence is assumed to be after the chickens. Actually, it has been shown that birds including game birds, song birds and poultry, make up less than 12 percent of the food of the red-tail hawk. The large size, light-colored underparts, black-tipped flight feathers, and a dark band across the body help identify the male red-tail in flight. As the name indicates, the upper surface of the tail in the adult is bright brick red and this may be seen as the bird wheels and turns or perches. Body length of the red-tail varies from 19 to 25 inches. The wingspread of the male is from 45 to 51 inches, and as in all the hawks, the female is larger. Her wingspread averages from 50 to 58 inches. The red-tail is a common resident in all areas of the state.

WATER OUZEL

The 'dipper' or water ouzel is a small, chunky slate-colored bird found along the banks of the Rogue. It is shaped like an oversized wren with a stubby tail. Legs are pale and eyelids are white. Note the bobbing motions when it walks. The uniqueness of this bird is in its ability to submerge beneath the surface of the river when feeding on small animals living in the mosses attached to submerged rocks. Some say it actually walks along the bottom. Zane Grey delighted in watching these unique birds, and he noted that he never saw two of them together.

Water Ouzel

CLIFF SWALLOW

These small birds can be seen flying over the water near their nests. The nest is a gourdlike 'jug' of mud, built on the faces of vertical cliffs or boulders. Note the birds' rusty or buff-colored rumps as identifying features. In flight, the swallow will glide, ending each glide with a much steeper climb, almost like the path of a 'roller-coaster.'

BELTED KINGFISHER

The kingfisher may be seen flying with uneven wingbeats, or hovering with rapidly beating wings in anticipation of a plunge to the water's surface. When the kingfisher is perched on a tree limb, you can observe its big head and large bill. A distinct white band resembles a collar around his neck. He is larger than a robin and has a brushy crest on top of his head. Note that when in flight he seems to always be cackling or rattling in a non-musical voice.

COMMON MERGANSER

The merganser is a very predictable year-round resident of the Rogue. When in flight a group of mergansers flies in formation, low over the surface of the water, along the winding course of the river. The males have a dark green head and long white body. The female is gray with a crested light red head. During the summer, the birds can be seen swimming with their young. Mergansers are one of the strongest swimmers in the duck family. Note that the merganser's bill is modified into a beak.

OTHER BIRDS

Watchful hikers or boaters might also see turkey-vultures, pileated woodpeckers, band-tailed pigeons, blue grouse, ruffed grouse, California quail, hummingbirds, varied thrushes, Oregon juncos, a variety of ducks and other miscellaneous migratory shore-birds.

Nancy Ross proudly displays her limit of steelhead.

FISH AND FISHING THE ROGUE

Fishing has made the Rogue River famous. Zane Grey's writings popularized the fantastic fishing potentials of this world-renowned river. Grey wrote much about the river and spent several summers and autumns camped along its banks, fishing for steelhead. Since then, fishermen from all over the world have come to the river to delight in casting a fly for this silver-sided beauty. Once hooked, the thrill of the fight will exceed your expectations.

STEELHEAD

The steelhead is a sea-run rainbow trout. It is the largest race of rainbow in Oregon. Although some steelhead remain in fresh water their entire lives, the Rogue steelhead migrates to the ocean during its first or second year, returning a year or two later when mature. Upon maturity it re-enters the Rogue to travel approximately 150 miles upstream to its spawning grounds. As the fish come into the river from the ocean, their size varies with the month of the year. The first steelhead begin to enter the river in the summer months of July or August and are quite small at that time. They may weigh only 1 or 2 pounds. This small-sized steelhead is commonly referred to as a 'half-pounder.' As the season progresses, the size of the fish increases. The winter-run steelhead are the largest, their size averaging some 6—7 pounds. During the months of January and February, the fish have reached their maximum size. The largest fish may exceed 15 pounds, but this is an uncommonly large size.

A female steelhead may deposit as many as 2,500 eggs, depending upon her size. Smaller females lay fewer eggs. These eggs are deposited in nests called 'redds'. The redds are 8—10 inches in depth and of varying widths. The female digs these redds with her tail and body. As soon as the eggs are laid, the male 'buck' swims slowly over the exposed eggs and deposits a milky fluid called 'milt' onto the surface of the nest to fertilize the eggs. Upon completion of the egg-laying ritual, the female 'doe' or 'hen' steelhead usually stays in the same general area for several weeks. Unlike salmon, the steelhead does not die after the spawn. Steelhead will return to the ocean, and about 10% of the fish will live to spawn a second time. The spring migration steelhead try to jump Rainie Falls, and this phenomenon can best be seen during the month of May.

The peak of the fly-fishing season runs from late August to mid-November, depending upon the weather and water conditions. During this time, thousands of fish are in the river starting their annual upstream migration to the spawning beds.

The best fishing for steelhead on the trip is usually found from the Battle Bar/Winkly Bar area down to John's Rapids. Also, when the river starts to widen, below Blossom Bar, there is excellent holding water and many steelhead accumulate in these calmer areas. Solitude Bar, Brushy Bar, Half Moon Bar and Tacoma Riffle are all excellent in productivity.

Actress Ginger Rogers and river guide Glen Wooldridge pose with freshly caught steelhead.

TROUT

Some of the larger tributaries on the lower river such as Big Windy Creek, Kelsey Creek and Missouri Creek offer good fishing for native trout. The fish will vary in size from year to year, depending on the flooding conditions of the previous winter. For example, after the great 1964 flood, most of the native trout were washed out of the fast-flowing mountain streams and the fishing the following summer was very poor. Best native-trout fishing in the tributary streams may be expected when at least two mild winters (with low-or no-flood conditions) have preceded.

SHAD

The shad is a member of the herring family, and was first introduced to the West Coast from the Atlantic Ocean in 1871. Since then it has established itself in our Rogue River with relative stability. Some years it is more frequently caught than others. The shad can be easily identified by its extremely compressed body and the five or six round black spots that appear on its sides. Shad rarely exceed 30 inches in length and normally weigh 2 to 6 pounds. Spawning occurs in late spring, generally in the third or fourth year of life. Late spring is usually the best time of year to try to catch this fish which is considered to be one of Oregon's sporting game fish.

The shad's fighting abilities are respected by sportsmen who refer to it as 'the poor man's tarpon.' Although extremely bony, shad is considered a delicacy when baked, and its roe is highy prized.

LAMPREY

The lamprey is easily identified by its long, slender, snake-like body. An average length is 12 inches, but it may reach two feet. Lamprey ascend the Rogue River to spawn over gravel beds in the headwaters of the river. Death occurs soon after spawning. Many dead lamprey may be seen during the summer months as they float downstream. The lamprey resembles an eel, but is not a true fish

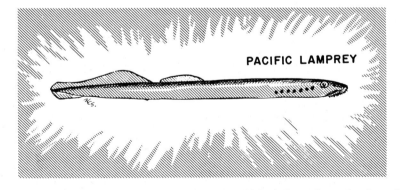

(like the eel), because his mouth is a sucker instead of jaw-structure, and his skeleton is made of cartilage instead of bone. Some anglers collect adult lamprey for sturgeon bait.

The chinook salmon is the second largest fish in the Rogue river. Truly a king among salmon, his great size and tremendous fighting qualities make him a fisherman's prize to be valued among the finest. The silver salmon also inhabits the Rogue, and is often confused with the chinook. Chinook can be positively identified by the black spotting on dorsal fins and on both lobes of the caudal (tail) fin. Silvers, however, have spotting on the upper half of the caudal fin only. Also, the inside of the chinook's mouth is all black, whereas the silver has a white gum- or tooth-line. The silvers weigh between 8 and 12 pounds when mature, however the chinooks range between 10 and 45 pounds in weight. The largest sport-caught chinook in Oregon was taken from the Umpqua River in 1910 and weighed 83 pounds! The Rogue has two spawning migrations (one in the spring and one in the autumn) however, you may find a few chinook in the river year-round. Spawning takes place only in the autumn months. Generally, spring-run chinook travel to the upper reaches of the Rogue and remain in deep pools until spawning time arrives. These magnificent fish can also be seen during the summer months as they school at the mouths of the numerous small creeks along the Rogue where water temperature is lower. The best locations to observe this schooling behaviour are at Rum Creek, Kelsey Creek, and Stair Creek. Unlike steelhead, all salmon die after they spawn.

The Rogue River salmon used to be fished commercially because its fine taste and firm quality meat commanded a very high price — in fact, in 1920 it is reported that some gill-netters earned over $1,000 in a single night's fishing (gill-netting was done in the dark so that the fish could not see the netting). Controversy developed between the gill-netters and the sport steelhead fishermen. Apparently the gill-netters were not only harvesting salmon, but also their nets were bringing in large steelhead. This decimation of the egg-producing steelhead and fine trophy fish upset the sportsmen considerably and resulted in open conflict among the fishermen of the river. This controversy motivated Zane Grey to write his novel 'Rogue River Feud' which focused upon the troubles between the two groups.

STURGEON

The white sturgeon inhabits the Rogue, in which it attains an enormous size. It is the largest fish found in fresh water in Oregon. A sturgeon may be as large as 1,000 pounds (several have been caught which weighed over 1,200 pounds!), and it may be as old as 80 or 90 years. This long, narrow-bodied fish, with a somewhat flattened head lurks at the bottoms of the deepest holes in the Rogue. It is slow growing seldom spawning before reaching an age of 15 years. The female deposits eggs in May or June. A 50 year-old fish may be depositing as many as 4,000 eggs.

If you're interested in catching this monster, your best chances are at Mule Creek Eddy (water depth exceeds 90 feet) and at the Sturgeon Hold just above Brushy Bar. Because of conservation efforts, only a specific size is permissible as a 'keeper' . . . to be sure, check current regulations before fishing for this rare species.

REPTILES OF THE ROGUE

WESTERN POND TURTLE

This turtle is commonly found along the banks of the Rogue. It usually ranges in length from 4″ to 8″ and has an olive-green/black color. Its legs are marked with red lines and distinct horizontal yellow lines near

its head and neck. When food is plentiful, its range is restricted to less than 100 yards. If removed any distance greater than this, the western pond turtle may be unable to return to its home. These harmless creatures are almost always seen in the quiet waters of the river, in back eddies and puddles. Water plants, aquatic insects, dragon fly larvae, snails, tadpoles and small fish make up the diet of the turtle. In the spring the female digs a hole in the sand or dirt and buries 7-10 eggs which hatch in the autumn. Turtles are mature at 4 years of

age when their size is 3½″ long. When you see them 'sunbathing' on a log or rock, it is for the purpose of drying out their skin to remove parasites. Like all cold blooded animals, the body temperature of the western pond turtle is dependent upon the environment, the warmer air conditions resulting in greater activity. In cold winter months, the turtles bury themselves in mud and remain protected until March or April, when they return to their riverside activities.

WESTERN FENCE LIZARD

The western fence lizard is the most common lizard in the United States. It may measure up to 10″ in length, of which 1/3 is tail length. Usually it has a smaller size and is colored gray with light brownish

wavy dark bands across the back. They spend most of their time basking in the sunshine. Males are very defensive of their own territory and frequent battles over domain rights often occur. This fighting leads fence lizards into a social hierarchy in which there is a distinct pecking

order with the No.2 'boss' male bossing all but No.1 male, and No.3 male bossing all but No.1 and No.2, etc. No. 1 male does almost all the mating with all the females in the territory. This insures that the population will remain vigorous and healthy. These interesting creatures can be found away from the moist river banks, in the rocky and drier areas of the Rogue canyon. the major food of the western fence lizard is insects and grubs of wood-boring beetles. The temperature of their environment controls the activity of the lizard, causing hibernation during colder winter months and resulting in a longer life span than southern cousins (in the south these lizards live less than 2 years because they do not hibernate, whereas in the Rogue canyon they may live to be as old as 8 years). Breeding occurs in the spring months. After mating, the female digs a hole in the ground and lays about a dozen eggs, which hatch in approximately 6-10 weeks (depending upon the weather) and the babies will be nearly 2″ long when hatched.

SKINK

This lizard is also commonly known as the 'blue racer' or 'blue-tailed racer' because of its most obvious characteristic, its bright blue tail. The skink may grow up to 12 inches in length. Its color is brown with pale longitudinal stripes running the length of its body. When young, the skink has a bright blue tail which acts as a safety device, because at this time of its life the skink is most vulnerable. Any predator will attack the bright blue tail because it is the most conspicuous object in sight. The tail breaks off and continues to wiggle as if it were alive itself. The predator becomes confused and the skink can shuffle safely away while the predator continues to attack the bouncing tail. The skink's streamlined scales and transparent eye coverings permit it to burrow easily into

young

adult

the shoreline sands. Some members of the skink family have limbs that are reduced or missing, which is an adaptation to the burrowing lifestyle followed by many of this species. The skink family contains over 600 species, of which many spend over half their lives underground. The main food of the skink is a diet of insects and other small animals, including grubs and larvae. About half of the 600 species lay eggs during reproduction, the other half bears their young live. Skinks may guard their eggs after they have been laid; also they may tend their eggs like mother hens, rotating the leathery sacks periodically.

WATER SNAKE

The name 'water snake' is likely to be given to any snake which spends most of its life in or near the water feeding on aquatic animals. The water snake can be easily distinguished from the rattler, because its head is the same width as its body and is rounded (whereas the rattler's head is broader than its body width and is

shaped like an arrowhead); also the water snake lacks rattles at the tip of its tail (almost all rattlers have at least one rattle attached to the tail), and the water snake seldom journeys more than a few feet from the riverbank (but the rattler can be found on high land and rocks many miles from the river). Water snakes are non-poisonous. When disturbed, the water snake escapes by diving into the water or by slithering under rocks or other debris along the bank. When picked up or alarmed they emit an evil smelling odor (which can be readily detected by smelling your hands after handling one). The water snakes of the Rogue canyon will seldom be found longer than 3 feet in length. The diet of water snakes consists largely of fish and amphibians, with the frog as the favorite on the menu. Water snakes bear their young live. An average of about 30 babies are born at a time.

VEGETATION ALONG THE ROGUE

The Rogue River Canyon was created by ancient lava flows, 140 million years ago. Pressures and high temperatures have altered the rock, folding it into a nearly vertical position in many places. Very little soil has been created over the past 100 million years because of the hardness of the rock structure of the canyon. In the area around Rainie Falls several fault zones can be seen. Shiny, greenish-black rock called 'serpentine' was squeezed like toothpaste into the weakened fault zone from deep in the earth. Soils that are formed from the serpentine erosion provide poor nourishment for plants to grow. Since few species can survive in serpentine soils the rock will be nearly bare. An example of of this can be seen at Tyee Rapids, where a solid rock ridge plunges into the river from the north bank. Plant growth is also restricted due to the rock slides caused by the steepness of the canyon. At Big Slide campground a landslide blocked the river for a time. At a point below Huggins Canyon the hard volcanic rock gives way to softer sedimentary rock. Better soil has therefore been created here and the powerful force of the river has carved a wider path through the canyon.

The amount of moisture is the most important factor controlling plant growth. The south bank of the river is exposed to less sunshine, due to the fact that the slopes are facing north. It is interesting to note that the north-facing slopes support a dense forest and understory, and that the same species found on the other side of the river are smaller and fewer in number. In fact, due to the difference in moisture content of the soil from one side of the river to the other, some species are only able to thrive on one side of the river. Not only will the size and density of a specific species vary from one side of the river to the other, the size of the leaves on the broadleaf species will tend to be larger on the south side of the river.

As the river flows toward the ocean the amount of rainfall increases. Agness has almost three times as much rainfall as Grants Pass. As you start your trip at Grave Creek, trees that require little moisture are found. Pacific madrone, white oak, canyon live oak, tan oak and manzanita inhabit the dry ridges. Downstream as the river leaves Mule Creek Canyon, large trees that require greater amounts of water are found. Large stands of Douglas-fir, hemlock, sugar pine and grand fir begin to appear.

WESTERN RED CEDAR

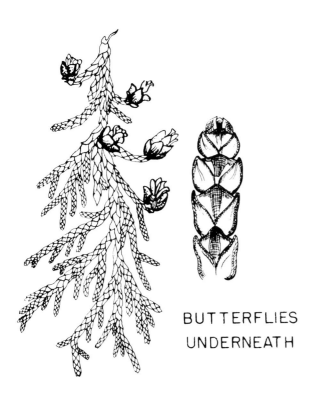

BUTTERFLIES
UNDERNEATH

Habit: large trees 150' to 200' high and 3' to 10' in diameter; with an open pyramidal crown of pendulous, frond-like branches. On large trees the base is fluted.

Leaves: 1/16" to 1/8" long (up to 1/4" long on older twigs), scale-like, in opposite pairs, closely pressed to the twig, white stomata pattern on the lower surface resembles a butterfly or bow tie.

Cones: about 1/2" long, erect, ovoid-cylindrical, scales are 10-12 in number but only 6 are fertile; all are in valve-like pairs.

Bark: 1/2" to 3/4" thick, fibrous; brown, but weathered to a grayish brown on the outside, finely ridged and furrowed; outer bark breaks up into long narrow strips or shreds.

Uses: early Indian tribes used this wood almost exclusively. Today it provides lumber for exterior siding, interior finish, greenhouse construction and flumes; boat-building, caskets, poles, posts, boxes and crates, sash and doors. 80% of the shingles and shakes manufactured in Oregon and Washington are made from western red cedar.

PORT ORFORD CEDAR

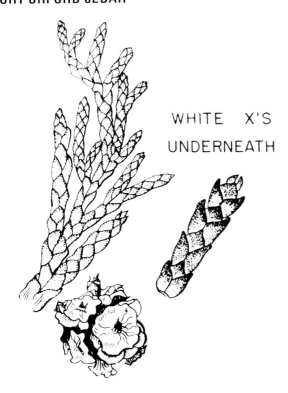

WHITE X'S
UNDERNEATH

Habit: trees are 125' to 200 ' tall, and 3' to 6' in diameter; with a pyramidal crown of pendulous, frond-like branches.

Leaves: 1/16" to 1/8" long, scale-like arranged in opposite pairs; mostly blue-green; glands on facial scales translucent when held up to the light; white X's on the underside of the foliage sprays.

Cones: round, average about 1/4" in diameter, reddish-brown with white powder under scales; scales are sprinkled on surface; mature in one season.

Uses: at one time the durable, easily-worked wood was commonly used for battery separators and venetian blinds; however, Douglas fir has replaced it for the former use and metal for the latter. A small volume is used each year for arrow shafts. Limited use as mothproofing for boxes and closets, interior trim, sash and doors, boat-building and dock planks.

DOUGLAS FIR

" PITCHFORKS "

(*Note: this is a common misnomer, actually the Douglas fir is not a real fir tree because its cones hang from the branches, rather than stand erect).

Habit: large tree 100′ to 250′ or more tall, 3′ to 6′ in diameter, with broad, pointed, pyramidal crown of dense foliage. *Needles:* about 1″ long, soft and slender. Linear and flattened with slight groove spirally arranged. Youngest branches often pendulous, especially on mature trees. *Cones:* 3″ to 4″ long, ovoid-cylindrical. *Bark:* on small trees gray or ashy-brown, thin, smooth, and with resiny blisters. On mature trees 3″ to 10″ or more thick near base, coarse, dark grayish brown, deeply and irregularly edged and fissured. Sloughing of the bark of very old trees may build up a mound around the bases of the trees. *Uses:* most important lumber tree in the nation. Structural lumber and timbers, veneer for plywood. Used for ties, poles, piling, battery separators, flooring, general construction, pulp and paper manufacture. Chemical derivatives of bark include tannin, waxes and dihydroquercetin (a food preservative). Over 20% of the saw timber volume in the United States is Douglas fir.

GRAND FIR (also called 'Lowland White Fir')

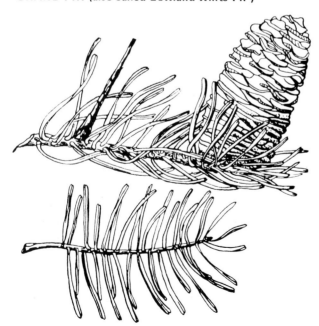

Habit: Large conifer, 125′ tall and 2′ to 6′ in diameter; with a long, narrow open crown that is rounded or flat-topped. *Leaves:* 3/4″ to 2″ long, lustrous dark yellow-green, grooved on upper side, 2-ranked. *Bark:* 2″ to 3″ thick on mature trees, furrowed and with flattened ridges; ashy brown in color and mottled with lighter colored areas, inner bark dark purplish-red. *Cones:* cylindrical, 2½″ to 4″ long, green to greenish-purple. *Uses:* General construction, boxes, crates, mill-work, pulpwood. *Remarks:* Grand fir seldom occurs in pure stands, but is usually in mixture with other conifers. It is more tolerant than Douglas fir and the pines, but less tolerant than the western hemlock, western red cedar and silver fir. It is commonly attacked by the Indian paint fungus.

WESTERN HEMLOCK

Habit: large trees 125' to 200 ' tall, and 2' to 4' in diameter. The pyramidal crown can be quickly identified because of the drooping center stem tip. *Leaves:* 1/4" to 3/4" long, linear, flat. Tend to be 2-ranked. *Cones:* 3/4" to 1" long, oblong, hang down, purplish-red becoming reddish-brown at maturity. *Bark:* thin, superficially scaly, brown to black on small trees; on old trees about 1" thick with flattened ridges; inner bark dark red streaked with purple. *Uses:* pulp, lumber for general construction, aircraft veneer, plywood. Bark is a source of tannin. *Remarks:* commonly associates with Douglas fir, western red cedar, western white pine, grand fir, silver fir, noble fir, mountain hemlock. Thin bark makes it very susceptible to logging and fire damage. This is the principal pulpwood species in our Pacific Northwest.

DROOPY TOP

SUGAR PINE

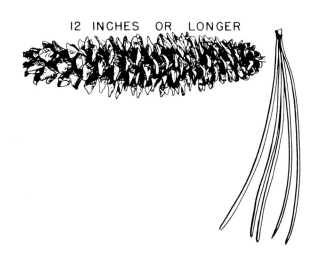

12 INCHES OR LONGER

Habit: largest of the pines, 150' to over 200' tall and 2' to 7' in diameter. Red, ridged bark. Base is free of branches for much of its length. Long cones hang from ends of branches. *Needles:* 5 per bundle, 2" to 4" long. *Cones:* 10" to 18", sometimes longer. *Seeds:* 1/2 to 5/8" long. *Bark:* thin, grayish-green and smooth on young trees; 1½ to 3" thick, with reddish, narrow, broken, scaly ridges separated by deep furrows on older trees. *Uses:* building construction, boxes and crates, sash, doors, blinds, interior and exterior trim, siding, panels, matches. *Remarks:* tree wounds secrete a sweet and sugary exudate which has cathartic properties. Seeds of sugar pine carried by Indians as emergency rations.

PONDEROSA PINE

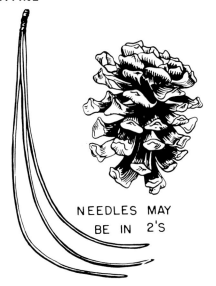

NEEDLES MAY BE IN 2'S

Habit: important large timber tree 125' to 180' tall, 3' to 6' in diameter. Yellow-brown bark in scaly plates. *Needles:* 3, sometimes 2, per bundle. Cones: 3" to 5" long (mostly 3" to 4"). Ovoid, green to purplish-brown prior to maturity, turning brown upon maturity. Armed with straight prickle. *Bark:* young bark is brown to nearly black, ridged and furrowed. Turns yellowish-brown in furrows near base of tree, gradually spreading over ridges and up trunk. Scales of bark look like jigsaw puzzle parts. *Uses:* millwork, boxes and crates, furniture, piling, poles, mine timbers and general construction. Probably the most important millwork and general-use species.

KNOBCONE PINE

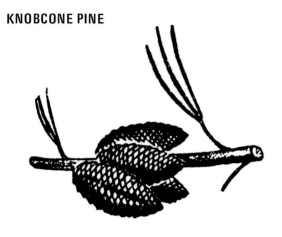

Habit: medium sized tree, 40' to 80' tall, up to 2' in diameter, usually of poor form. *Needles:* 3 per bundle, 3" to 7" long, slender and flexible. *Cones:* 4" to 6" long, ovid-conic, yellowish-brown, asymmetrical at base. Cones are recurved on twigs, 3 to 6 in a cluster. They persist indefinitely on the tree unopened. Will open following a fire. *Bark:* thin gray-orange-brown or gray-reddish-brown on upper trunk and larger limbs. Small, flaky scales. *Uses:* only common use is as a local fuel.

BREWER SPRUCE (also called 'Weeping Spruce')

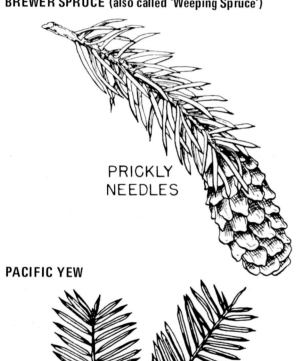

PRICKLY
NEEDLES

Habit: has a sparse, open crown and long pendulous middle and lower branches. 50' to 80' tall, 1½ to 2½' in diameter. *Needles:* 3/4" to 1" long, tend to point forward on twig. *Cones:* 2½" to 6" long, oblong-cylindrical, with rounded scale-tips, purplish-red when young, reddish-brown at maturity. *Bark:* less than 1" thick, reddish-brown with long, firmly attached scales. *Remarks:* the rarest and least known of American Spruce. Difficult to propagate.

PACIFIC YEW

CONIFEROUS
UNDER—TREE

Habit: small trees 30' to 50' (sometimes up to 80' tall,) and 1' to 2' in diameter, with a dark green crown of pendulous branches. Usually a moist-site, understory tree. *Leaves:* 1/2" to 1" long, linear, rigid, dark green above, paler beneath; commonly 2-ranked. *Fruit:* fleshy orange-red, 3/8" to 5/8" long, oblong-oval. *Bark:* about 1/4" thick, dark purplish or reddish-brown, scaly; inner bark reddish-purple. *Uses.* The wood is very durable and very hard; it requires no preservative treatment even when in contact with the soil. Used for special fence posts, gate posts and corner posts. Makes good archery bows.

RED ALDER

Habit: moderately open, broadly pyramidal or dome-shaped crown; 30' to 120' tall and 1' to 3' diameter; *Leaves:* 3" to 6" long and about ½ as wide, ovate to ovate-elliptical. Shiny green to yellow-green and smooth above. *Fruit:* tiny nutlets with thin lateral wings or one encircling wing, ½" to 1" long, cylindrical brown to gray. *Bark:* ashy gray to grayish brown, smooth along upper bole and near base with flattened irregularly plated ridges which are superficially scaly; inner bark tan, becoming reddish brown upon exposure to the air. *Uses:* furniture (often disguised as maple or some other species), core stock and cross bands in plywood, woodenware, millwork, plugs for paper rolls, and limited usage for pulp. *Remarks:* forms pure stands or occurs in mixture with cottonwood, bigleaf maple, vine maple, Oregon ash, willows, Douglas fir and grand fir. Red alder is the largest and most important of the Pacific Coast alders. The total volume of this species is greater than any other western hardwood.

OREGON ASH

Habit: tree 40' to 80' tall and 1' to 2½' in diameter with narrow or broad crown. *Leaves:* 5" to 14" long. Light green and smooth above, paler below. *Flowers:* small, white, borne in dense clusters. *Fruit:* single, winged fruit, 1½" to 2" long. *Bark:* up to 1½" thick, dark gray or gray brown, furrowed and with flat ridges. *Remarks:* usually associated with bigleaf maple, red alder, willows, Oregon white oak, Douglas-fir and grand-fir. Used for shovel, rake and hoe handles, furniture, baseball bats, oars, baskets, boxes and crates, boat-building and cooperage. Fairly important browse for deer and elk.

WESTERN AZALEA

Habit: loosely branched shrub up to 10' high. *Leaves:* deciduous, 1½" to 4" long, ½" to 1" wide, elliptical. Green and smooth above; paler and smooth beneath. *Flowers:* white or tinged with pink, 1½" to to 1¾" long, borne in loose clusters. *Fruit:* capsule about ½" long, brown. *Remarks:* poisonous to livestock. Planted as an ornamental.

EVERGREEN BLACKBERRY

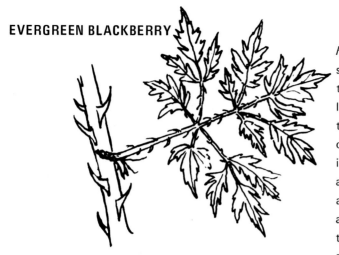

Habit: shrubs with heavy, angular, trailing or climbing stems up to 10" in length. Armed with large, flattened, recurved prickles. *Leaves:* 3 to 5 lacerated leaflets; green to reddish-green above. Flowers: white to pink in color. *Fruit:* black aggregate of small drupelets. *Remarks:* commonly found on barren and infertile soils on burns, old fields, logged-over areas and along roadsides. It was introduced from Europe and escaped from cultivation. Fruit is made into jams and jellies. Plant has little forage value. The dense thickets provide food and excellent cover for birds and small mammals.

CASCARA BUCKTHORN

Habit: 50' tree, 10" to 20" in diameter, sometimes an erect shrub up to 15' high. *Leaves:* 2" to 6" long, oblong, dark glossy green above, paler below; thin to somewhat leathery in texture. *Flowers:* small, green-white, borne in loose clusters. *Fruit:* round, ¼" to ½" in diameter, black on the outside with a yellowish sweetish pulp. *Bark:* thin, grayish brown to gray-reddish brown, smooth or somewhat scaly (on large trees), often mottled with chalky-white patches. Inner bark yellow. Tastes bitter. *Remarks:* the cured bark is one of the most important natural drugs produced in North America, and is commonly used as a laxative. The species seldom reaches a very large size due to the stripping of the bark. If a tree's bark has been incorrectly stripped, the tree should be felled so that the stumps will sprout. Birds and some of the smaller mammals will eat the fruit. Deer may crop the foliage and twigs.

GOLDEN CHINKAPIN (also called 'Western Chinquapin')

Habit: evergreen trees 90' to 150' in height, and 3' to 6' in diameter, with round or conical crown. *Leaves:* 2½" to 4½" long, oblong-elliptical, leathery, pale green or yellow green. *Flowers:* borne in erect catkins. *Fruit:* 1 or 2 somewhat triangular nuts, borne in a 4-parted burr with sharp, branched spikes, light brown when mature after 2 growing seasons. *Bark:* on young trees grayish brown and mottled with large white areas; on older trees broad, flat, dark ridges, deeply furrowed, 1" to 2" thick. *Uses:* limited use as construction lumber boxes and crates, furniture and cabinet work. *Remarks:* closely related to the chestnut species, it is susceptible to chestnut blight, the disease which has practically eliminated the American chestnut in the East.

PACIFIC DOGWOOD

Habit: trees up to 60' tall (usually much smaller), with a round to oblong-conical crown. Leaves: 3" to 5" long, 1½" to 3" wide, broadly elliptical to ovate; bright green above, paler below. Leaves red in autumn. *Flowers:* very small, greenish white, borne in a dense compact head and surrounded by 4 to 6 broad, creamy white bracts. *Fruit:* flattened, reddish fruit borne in a tight cluster. *Bark:* thin, dull gray, smooth. *Remarks:* often found as an understory tree. While the tree may flower in heavy shade, it will not bear fruit. Associated with bigleaf maple, red and white alders, vine maple, willow, Douglas-fir, western hemlock and redwood.

WESTERN DOGWOOD (also called 'Creek Dogwood' or simply 'Dogwood')

Habit: a large, loosely branched shrub with reddish stems, up to 15' high. *Leaves:* 2" to 6" long, ovate, dark green and lustrous above, pale gray below. Leaves turn red in the fall. *Flowers:* small, white, borne in flat-topped, terminal clusters. *Fruit:* white or ivory-colored, berry-like, about ¼" in diameter. *Remarks:* foliage and new twigs browsed by deer.

CALIFORNIA HAZEL

Habit: open and spreading shrubs up to 15' tall, or small trees up to 30' tall and 6" to 12" in diameter. *Leaves:* broadly ovate, 2" to 4" long and 1½" to 3" broad, margins sharply or doubly serrate. *Fruit:* nuts about 3/8" to 5/8" long, wholly surrounded by tan-colored, somewhat hairy, papery husk. *Remarks:* tolerant to many environments, nuts edible.

HIMALAYA BERRY

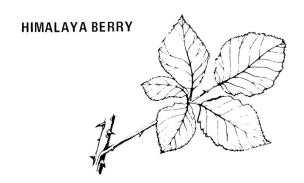

Habit: an erect, spreading or trailing shrub, with stout heavily ridged and armed stems. *Leaves:* 3 to 5 oval leaflets. 1½" to 2½" long, dark green to reddish-green and smooth above. *Flowers:* large, white, borne in clusters. *Fruit:* a black aggregate of drupelets some 1" in length. Used for jams and jellies.

EVERGREEN HUCKLEBERRY

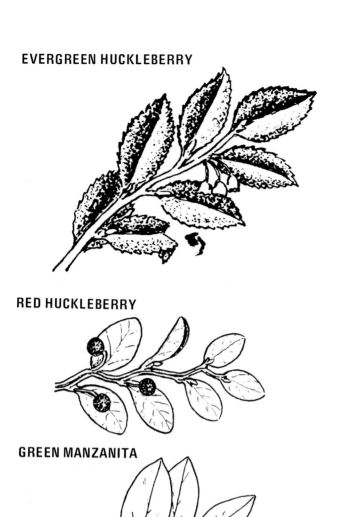

RED HUCKLEBERRY

GREEN MANZANITA

PACIFIC MADRONE

Habit: evergreen shrub, with small, glossy, waxy leaves; up to 10' tall. *Leaves:* ½'' to 1½'' long, ovate or oblong; thick and leathery; dark glossy green. *Flowers:* pink, about ¼'' long, borne in small clusters. *Fruit:* round, bluish-black berry, up to ¼'' in diameter. *Remarks:* often associated with salal, hairy manzanita, rhododendrons, Pacific waxmyrtle, Sitka spruce, western hemlock and Douglas fir. Fruits are eaten by birds, mammals and man. Foliage is used for greenery by the florists. Ornamental shrub. Important browse for elk and deer.

Habit: a shrub 4' to 10' tall (occasionally to 20'). *Leaves:* ½'' to 1½'' long (generally under 1''); dark green above, paler beneath. *Flowers:* small, greenish-white or sometimes reddish. *Fruit:* a bright red berry with a translucent skin, about ¼'' in diameter. *Remarks:* berries are palatable to birds, mammals and humans.

Habit: evergreen shrub up to 6' in height. *Leaves:* 1'' to 2'' long, ovate to elliptical; light green and glossy on both surfaces. *Flowers:* pinkish-white, urn-shaped, borne few to a cluster. *Fruit:* round, chestnut-brown to black, berry-like fruit about ¼'' in diameter. *Bark:* smooth reddish-brown, exfoliating. *Remarks:* a good browse for mule deer.

Habit: evergreen tree 60' to 100' tall and 2' to 6' in diameter, with sloughing bark. *Leaves:* 3'' to 5'' long, 1½'' to 3'' wide, leathery, oblong, light green when they first unfold, becoming dark green above, pale silvery green below. Dead leaves fall in spring or early summer. Leaves stay green 13 or 14 months. *Flowers:* white, urn-shaped, about ¼'' long, borne in clusters up to 6'' long. *Fruit:* orange-red, pebbly-skinned, berry-like fruit about 1/3'' in diameter. *Bark:* on young stems, thin, red or orange-brown, separates into scales or short strips and exfoliates; on large trees, reddish brown, scaly and flaking. *Uses:* little used. Has been used to manufacture charcoal, and can be used for furniture and paneling. Substituted for Dogwood in shuttles. Wood is difficult to dry because of its great tendency to warp and check.

HAIRY MANZANITA

BIGLEAF MAPLE

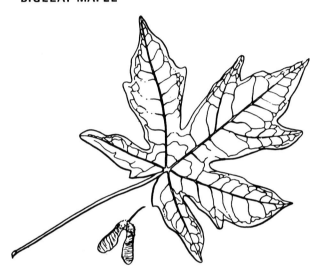

VINE MAPLE

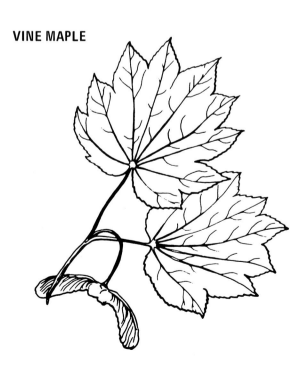

Habit: bushy shrub or small tree, with stiff and somewhat gnarled branches — seldom exceeds 12' in height and 5" in trunk diameter. *Leaves:* leathery, 1" to 2" long, elliptical, round or oval; dull green or pale blue-green, hairy on both surfaces. *Flowers:* pinkish-white, urn-shaped, borne in small terminal clusters. *Fruit:* smooth, red, berry-like, 1/4" to 1/3" in diameter. *Bark:* smooth, brown on young stems, exfoliates, revealing light-colored underbark. *Remarks:* 'Manzanita' is Spanish for 'little apple.' An attractive shrub which leaves a messy accumulation of bark at its base.

Habit: tree may be 40' to 100' tall and 2' to 4' in diameter; when growing in the open, it usually branches within the first 15 feet into several large branches. *Leaves:* 6" to 12" in diameter; 5-lobed, central lobe usually wedge-shaped and narrow-waisted. *Flowers:* small, yellow, borne in long racemes just ahead of the leaves. *Fruit:* double winged structures 1½" to 2" long. *Bark:* smooth, grayish brown on young trees; grayish brown to reddish brown on old trunks. *Uses:* furniture and paneling, flooring. Burls weighing from a few hundred pounds to several tons are cut and shipped to France and Italy where they are sliced into veneer to be used in furniture manufacture. *Remarks:* good browse. Bigleaf maple has the largest leaves of any of the maples. The sap has a high sugar content, but weather conditions in the Rogue area are not conducive to a high flow of sap.

Habit: an erect shrub or more commonly a helter-skelter arrangement of crooked branches that are a curse to anyone who has need of passing through them; up to 20' tall, or less commonly a small tree 30' to 40' in height. *Leaves:* circular in outline, averaging 2" to 4" in diameter, or sometimes slightly larger. If growing in the sun, the leaves may take on a red color early in the summer. *Flowers:* red, borne in short, terminal clusters. *Fruit:* a propeller-like double-winged structure. *Remarks:* common understory species, also found on cutover and burned over lands. Valuable forage for deer and elk. May be used locally for fuel. Effective for smoking fish and fowl. Larger stems used to fasten logs together in building on-site rafts. Indians used the branches for bows. Very colorful in the autumn.

115

OREGON MYRTLE(also called 'Bay', or 'California Laurel')

Habit: large, evergreen trees 60' to 100' tall and 2' to 5' in diameter, with aromatic foliage; or a prostrate to erect shrub up to 15' high. *Leaves:* aromatic, 2½" to 5½" long and up to 1" wide, elliptical to oblong, dark green, very strong-scented when crushed. *Flowers:* inconspicuous, yellowish. *Fruit:* bluish-black, olive-like, about 3/4" in diameter. *Bark:* on young trees smooth and dull grayish brown. *Uses:* turnery items, novelties, veneer, furniture, cabinet work, keel blocks and friction blocks. *Remarks:* vigorous sprouter. Aromatic leaves and volatile oils will irritate the eyes and nose.

CALIFORNIA BLACK OAK

Habit: trees 40' to 80' tall and 1' to 2½' in diameter, with open, rounded crown. *Leaves:* 3" to 6" long, 2" to 4" wide, oblong to obovate. Margins mostly 7-lobed, lobes 3-toothed and bristle-tipped. *Fruit:* nut (acorn) long, 1" to 1½", chestnut-brown, requires two growing seasons to mature. *Bark:* dark gray or black and smooth on young trees, dark brown with reddish tinge on most old trees. About 1" thick. *Remarks:* occurs as a scattered tree or in open, pure stands, but more commonly associated with Douglas fir, incense-cedar, Oregon white oak and canyon live oak. Deer feed on leaves and acorns.

CANYON LIVE OAK

Habit: evergreen trees 30' to 80' tall and 1' to 2' in diameter, with a dense, broad, round crown, or also occurring as a shrub up to 15' tall. *Leaves:* persistent, 1" to 3½" long and ½" wide, elliptical, thick and leathery; fuzzy when first appearance made, becoming yellow-green, lustrous and smooth above. Margins have holly-like spinose teeth. Species has 2 types of leaves and it is common to find both types on the same plant, even on the same twig. *Fruit:* nut (acorn) ovoid to oblong, ½" to 2" long, and about ½ as wide. Initially somewhat fuzzy. Requires 2 seasons to mature. *Bark:* grayish-brown tinged with red, mostly smooth, with small closely appressed scales, 3/4" to 1½" thick. *Remarks:* tolerant, occurs in pure small stands, but commonly associated with Douglas fir, incense-cedar, various live oaks. Vigorous sprouter. The species is considered to be the most ancient of the existing American oaks. The wood has only limited use, it is heavy, hard and strong.

OREGON WHITE OAK

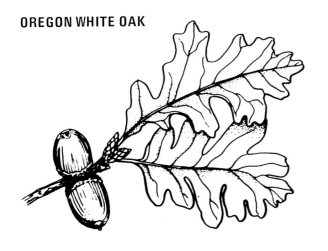

Habit: trees 40′ to 80′ tall and 2′ to 3′ in diameter, with a broad, compact crown. *Leaves:* deciduous, 3″ to 6″ long, 2″ to 4½″ broad, oblong, thick and leathery; dark green; margins 7-to-9-lobed, lobes round or bluntly pointed. *Fruit:* nut (acorn) oval or barrel shaped, ¾″ to 1½″ long, half to 2/3 as wide. Matures in one season. *Bark:* white to light brown or grayish-brown, shaggy or with short, broad ridges and shallow furrows, less than 1″ thick. *Uses:* fuel; potentially valuable for flooring, furniture, cooperage, cabinet work, interior trim and ship-building.

OCEAN SPRAY (also called 'Arrowwood')

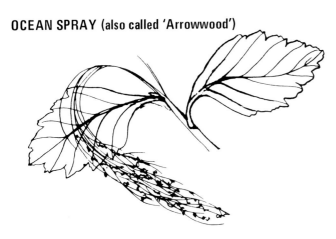

Habit: erect, loosely branched shrub up to 15′ tall. *Leaves:* ¾″ to 2½″ long, ovate, green, margins coarsely toothed or very shallowly lobed. *Flowers:* very small, white or creamy, borne in dense clusters. *Fruit:* light brown, tiny, 1-seeded follicle; fruit clusters persist into the winter, or until the next growing season. *Remarks:* browsed by elk and deer. The Indians used the straight stems for arrow shafts, hence one of the common names for the species 'Arrowwood.'

DWARF OREGON GRAPE

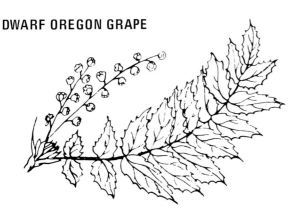

Habit: low, evergreen shrub with spine-tipped leaves. *Leaves:* 10″ to 16″ long, ovate, dark glossy green above, paler beneath and without a distinct midrib. *Fruit:* dark blue berries about 3/16″ diameter. *Remarks:* common in the understory along the Rogue. Florists use the foliage for greenery. Birds and small mammals eat the fruit.

TALL OREGON GRAPE

Habit: erect, evergreen shrubs 3′ to 10′ tall, with dark green glossy leaves. *Leaves:* 6″ to 12″ long, dark glossy green above, paler beneath. Spiny tips. *Fruit:* dark blue berries about 3/16″ diameter. *Remarks:* state flower of Oregon. Fruit is eaten by many birds and mammals. Makes good jelly. Nurserymen and florists use the foliage.

POPLAR (also called 'Cottonwood')

Habit: rapidly growing trees. May be tall, narrow, up to 150' in height, or shorter, broader, up to 75' tall. *Leaves:* deciduous, alternate, simple, 5-ranked. Upper surface smooth and glossy. Deltoid or rhomboid in shape. Margins may be dentate, serrate or lobed. *Flowers:* borne in drooping catkins, appearing before the leaves. *Fruit:* 2-to 4- valved, 1-celled capsule; ovate or conical in shape; usually less than 1/4'' long, smooth or with very fine hair. Contains several minute, hairy seeds. Fruits mature in late spring. *Uses:* wood used for pulp, boxes and crates, excelsior, veneer, matches and factory lumber. *Remarks:* 15 species are native to the United States. Very intolerant, fast growing, moisture demanding. Short-lived, but vigorous sprouters. The litter caused by fallen flowers, the obnoxious quality of the cottony seed and the tendency of the roots to break through sewer joints and clog the lines, militates against the use of the poplar for ornamental planting in populated areas.

WESTERN RASPBERRY (also called 'Blackcap')

LEAVES

WHITE

BENEATH

STEMS HAVE WHITE POWDER

Habit: a semi-erect shrub with round, armed trailing stems. *Leaves:* 3 to 5 ovate leaflets, dark green. *Flowers:* white, borne in small clusters. *Fruit:* red, dark purple or black aggregate of drupelets, about ½'' in diameter. *Remarks:* the fruit is eaten by humans, birds and mammals. It is a fair browse plant.

PACIFIC RHODODENDRON

Habit: a straggly, evergreen shrub up to 12'' high. *Leaves:* 3'' to 6'' long, oblong, thick and leathery. Dark green and smooth above, paler below. *Flowers:* rose-purple, occasionally white, 1'' to 1½'' long, borne in a round, loose cluster. *Fruit:* capsule about ½'' long, brown in color. *Remarks:* fruits are gathered and sold commercially. Numerous birds and mammals feed on the fruit. Foliage and younger twigs are an important game browse in some localities.

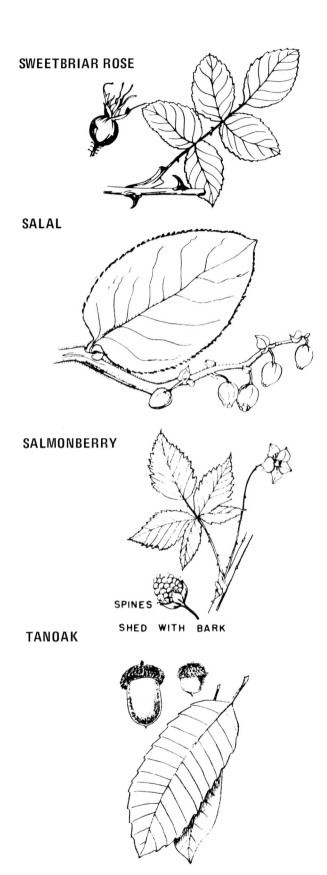

SWEETBRIAR ROSE

SALAL

SALMONBERRY

SPINES
SHED WITH BARK

TANOAK

Habit: erect shrub up to 8' tall; stems armed with coarse, recurved prickles. *Leaves:* 1½" to 4" long, with 5 to 7 oval or oblong-oval leaflets; dark green and smooth above, pale below. Emit a sweet 'cider-like' odor when bruised. *Flowers:* mostly pink, occasionally white. *Fruit:* hair covering colored red, smooth, ovid to ellipse-shaped 'hip' berry which is ½" to ¾" long.

Habit: an evergreen shrub, 1' to 2½' in height (occasionally taller). *Leaves:* 1½" to 3½" long, ovate, leathery, dark glossy green and smooth above, paler beneath. *Flowers:* Pinkish, urn-shaped, about ¼" long, borne in loose clusters. *Fruit:* bluish-black, ovoid berry about 5/16" in diameter. *Remarks:* Salal is perhaps the most common shrub in the understory of the Rogue forests. Many birds and mammals feed upon the fruit. Deer and elk will browse salal occasionally. The foliage is used for greenery by florists.

Habit: an erect, moist-site shrub 3' to 10' high, with light-brown, peeling bark. *Leaves:* 1" to 3" long, shiny dark green and smooth to wrinkled above, paler beneath. *Flowers:* pink to dark red, borne singly. *Fruit:* an aggregate of drupelets; salmon colored to red or reddish-purple. *Remarks:* provides food and cover for birds and small mammals. The watery, poorly flavored fruit is eaten, but seldom gathered by humans.

Habit: evergreen trees 60' to 100' tall and 1' to 3' in diameter, with a dense, broad, round crown, or also occuring as a shrub up to 10' in height. *Leaves:* 3" to 5" long, thick, stiff and leathery, oblong to ovate, dark green at first, later pale green; lustrous and smooth. *Fruit:* nuts (acorns) oval to ovate, borne singly or paired, ¾" to 1¼" long. *Bark:* ¾" to 1½" thick, with narrow furrows and broad, rounded or flattened ridges checkered with square plates which are superficially scaly, reddish brown to grayish brown in color. *Uses:* locally for fuel, furniture and mine timbers. Bark is a commercial source of tannin. *Remarks:* occurs in small, pure stand. Birds and rodents feed on the acorns. Vigorous stump sprouter.

119

THIMBLEBERRY

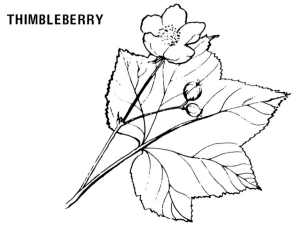

Habit: an erect shrub 3' to 6' high, with weak, cane-like unarmed stems. *Leaves:* deciduous, simple, broad 3" to 8" in diameter. Entire leaf is soft to the touch. *Flowers:* white or whitish-pink. *Fruit:* red, flattened hemispherical, easily pull free. *Remarks:* a fair to outstanding browse plant. The fruit is eaten by birds, mammals and humans.

WAXMYRTLE (also called 'Bayberry')

Habit: large, evergreen shrub or small tree up to 30' or 40' tall, and 8" to 12" in diameter, with a dense dark green, round crown of slender "willow-like" foliage. *Leaves:* alternate, simple, 2" to 4" long, 1/2 to 3/4" wide; dark green, smooth and shiny above, paler beneath. *Flowers:* catkins borne in leaf axils. *Fruit:* round, dark purple to grayish-white, waxy. *Bark:* very thin, dark gray or grayish-brown and often mottled with white areas. *Remarks:* very tolerant. Associated with evergreen huckleberry, rhododendron, salal, hairy manzanita, Sitka spruce and western hemlock. Planted as an ornamental. Early settlers gathered the fruits and rendered the wax to make candles.

WILLOWS

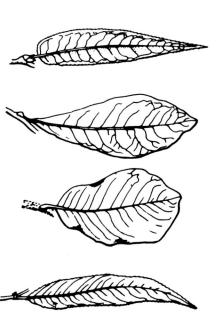

Habit: rapidly growing, thicket-forming trees and shrubs. *Leaves:* deciduous, alternate, 5-ranked, margins may be entire, wavy or serrate. *Flowers:* forms catkins, sometimes fragrant. *Fruit:* two-valved, 1-celled capsule, ¼" or less in length; contains several minute, hairy seeds. Seeds disseminate in late spring or early summer. Seed very short-lived. Needs moist mineral soil. *Remarks:* very intolerant. Occurs on many types of soil. Comparatively short-lived, but prolific sprouters. Easily propagated by cuttings. Useful for erosion control.

DIFFERENT SPECIES

INTRODUCTION TO FERNS OF THE ROGUE

For every tree there are 100 ferns. Bracken, sword, maidenhair, deer and licorice fern may be easily found and identified along the Rogue. Ferns are favorites for decorative uses, both in the garden and in the florist's art. The Rogue's sword fern has both beauty and keeping qualities, making it known to florists everywhere.

Legends and fallacies about ferns have developed because our ancestors could not understand where they came from. No seed could be found, but, since all plants were supposed to grow from seeds, some thought the 'seed' must be invisible. It was thought that the person finding the 'seed' of a fern would gain magical powers, such as enabling himself to become invisible. Although modern biologists have identified the sori and sporangia and clarified the reproductive cycle of the fern, some of the romance of days gone by still lingers in our families' folk-tales.

Besides the beauty and mystery of the ferns, there is a practical side. Bracken fern rootstock has an abundance of starch the year round and therefore is a possible source of food for a person lost in the woods. The tender new tips of bracken fern may also be boiled and flavored with sugar, pepper, salt, garlic, catsup, sesame or other spices and eaten as a naturally-occurring delicacy.

Ferns were once mighty plants, large as trees and thick as the forests. That was 40 million years ago, when the climate of the Rogue area was hot and moist, and when great dinosaurs were making tracks in the mud — tracks that are found today in Eastern Oregon. Geological changes pressed down the tree ferns and they became coal and oil. Today's machine age is powered by fuel produced by the ancient ferns.

BRACKEN FERN

SWORD FERN

MAIDENHAIR FERN

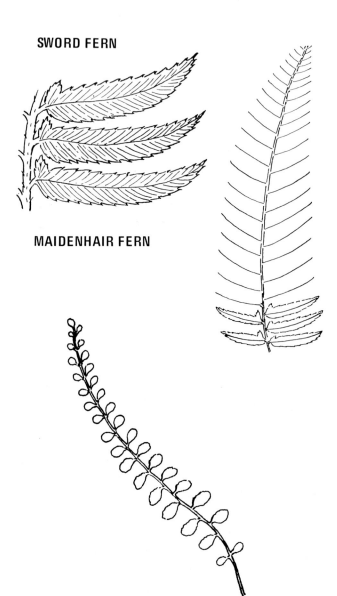

Everybody recognizes this common fern. It tends to form dense cover, usually 2' to 6' high. Bracken often overwhelms vacant fields, thereby becoming an annoying weed. It thrives in both dampness and dryness, and is ideally suited for the Rogue canyon. The plant has also commonly been called 'brake', which traces back to its European history, possibly because of the broken appearance of the fern after the first heavy frost of each year, when its leaves collapse. Indians dug up the underground rootstocks, roasted them and extracted a starchy substance which they used as food. The young shoots may be boiled and eaten as a vegetable. Hunters and campers often use the Bracken Fern to prepare a soft, resilient bedding base.

The leaves of the sword fern are evergreen and may reach a height of 4 feet or more, but are commonly 1' to 3' high. The slim, curving leaves rise from large, circular clusters. Each leaf carries dozens of alternating leaflets, each of which resembles a short broad sword, giving the plant its name. This tough, coarse fern is found particularly in shaded parts of the Rogue valley, near moisture-laden soils. Sprays and wreaths are prepared from this fern by the florists around the country.

This delicate fern is easily recognized because of its pleasing appearance, unlike that of any other fern. Its erect, black, shiny 'wire' leafstalks may have suggested the name, or the name may refer to the fine, black, fibrous roots. Each tiny leaf division is fringed upon the upper edge. The leaf blades are soft and thin. Notice the pattern of characteristic concentric circles evident in the picture. The fern prefers the dripping banks of the streams along the Rogue. It commonly grows 1 foot high and is usually not over 2 feet. The plant's scientific name is from the Greek meaning 'not to moisten.' The leaves resist water, or are so smooth that water runs off readily. The Roman naturalist and writer Pliny says, " In vain you plunge it in water, you cannot wet it." Indians of the Rogue used the thin leaves to dry berries. Over a fire they placed a lattice of cedar strips. Leaves of fern were strewn on the strips and then berries were spread on the fern. The dried fern leaves were winnowed out of the dried berries.

LADY FERN

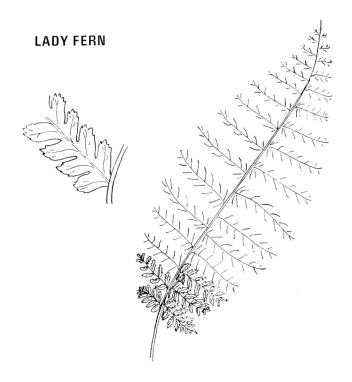

Lady fern grows in damp, shady woods alongside the numerous creeks entering the Rogue. It is commonly 2' to 5' high. The plants are not evergreen. By autumn they often look injured and the leaves lose their clear green color. This graceful fern is easy to identify by its leaves, as they are widest near the middle and taper evenly toward top and bottom. Some have compared the edges of the fern's leaves as having long, sweeping curves like giant parentheses. Sir Walter Scott said of the attractive lady fern:

"Where the copse wood is the greenest,
Where the fountain glistens sheenest,
Where the morning dew lies longest,
There the lady fern grows strongest."

DEER FERN

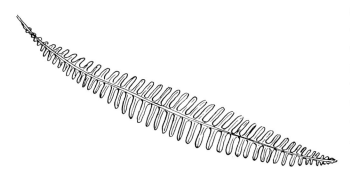

Commonly grows 1' to 2' high in the deep shade of the Rogue canyon. It likes wet soil and associates with the sword fern. Leaves grow in a close tuft or rosette. Its name probably comes from the fact that it is a source of food for deer and elk during winter snows, when they may be seen pawing away the surface whiteness to reach the resistant blades of fern underneath.

LICORICE FERN

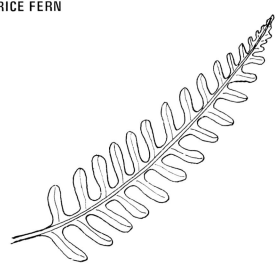

Almost any fern growing out of thick moss on tree trunks or rocks will be of this genus. Particularly likely to be found on the lower trunk and limbs of bigleaf maple trees, if they have been padded with moss. Commonly not over a foot long, the licorice fern is sparsely found. In times of little or no rain, the leaves wither, but after the rainy season begins they reappear and the fern remains green all winter. Early settlers used the rootstocks to flavor tobacco and to make remedies for a variety of ailments. The rootstocks, usually found under the moss on trees and rocks, contain the licorice flavor.

Author holds rattlesnake. Note venomous fangs.

HAZARDOUS PLANTS AND ANIMALS ALONG THE ROGUE

POISON OAK

SHINY LEAVES

Habit: erect shrubs 3' to 10' in height or tree-climbing vines with shiny, dark green leaves which turn varying shades of red and yellow in the fall. *Leaves:* alternate, 3 ovate shaped leaflets. Terminal leaflet is larger than the 2 lateral leaflets. Surfaces smooth and shiny. *Flowers:* small, long-stemmed, inconspicuous, yellowish green, borne in loose, pendulous clusters. *Fruit:* round to globular, grayish white, striated, persist after the leaves have fallen. *Remarks:* occurs on many of the slopes and camping sites along the Rogue. Poison oak is toxic to most individuals, immunity is a relative subject. Individuals who have not been bothered by poison oak for some years may find themselves afflicted. Fumes from the burning plants are especially toxic, causing respiratory reactions which may be extremely critical. Bees are attracted to the flowers in the spring, but none of the toxicity of the plant is transmitted through the nectar. Horses and cattle can browse the species with immunity. Household pets that have wandered through the plant are carriers of the plant's irritating resins.

RATTLESNAKE

These are heavily-bodied, highly venomous snakes. When disturbed, the rattler vibrates his tail (and rattle) as if giving warning that he is about to strike. The rattle is made up of a number of loosely interlocked shells, each of which was the scale originally covering the tip of the tail. The end scales tend to wear out from use, so there can be a different number of segments in the rattles of two rattlers of the same age. In the wilds, these rattles seldom exceed 14 in number. The most effective number is 8, larger rattles resulting in muted sounds. Unless provoked, the Timber Rattler of the Rogue River Canyon will make no attempt to strike. This is unlike the Western Diamondback Rattler, which may pursue an intruder. How poisonous a snake is depends upon several factors: its age (the younger it is, the less the amount of venom); its natural range (snakes of the same species from one part of a range may be more venomous than from another part of the same range). Once depleted, the venom may take as long as 2 months to regenerate completely. The main food of the rattlesnake is small

warm-blooded animals, such as mice, chipmunks and other small rodents. Young rattlesnakes feed mainly on cold-blooded animals such as frogs, salamanders and lizards. A rattlesnake requires only 1/10 as much water as other mammals of comparable size and weight because its skin is almost perfectly waterproof. All rattlesnakes give birth to live young. Mating is in the spring, and the litter may vary from 1 to 60 baby snakes, depending upon the size of the mother (the usual number is from 10 to 20). Few baby rattlesnakes survive their first year because of the high toll taken by predators and enemies, such as hawks, skunks and snake-eating snakes; pigs, deer and other hoofed animals. Watch for rattlesnakes during the heat of the day, at which time they will be located in the shade or rocks, trees and plants. In the evenings and early mornings they will come out into the open to sun themselves on the sands and open rocks. Expect to see rattlesnakes any-where along your journey down the Rogue. Be prepared, carry a snake-bite kit and be familiar with its correct use.

SCORPION

In the United States it has been estimated that more people are killed annually by scorpions than by snakes. Scorpions are notorious for their stings, the venom of which is sometimes fatal to man. They range in length from ¼" to sometimes as long as 8". The species found along the Rogue River Canyon seldom exceeds 2". The body is segmented and the head bears a pair of pincer-like claws, similar to the lobster. The thorax has 4 segments, each with a pair of walking legs. The abdomen has 6 segments, tapering to a single sharp sting at the end with a small opening supplied by two relatively large venom glands. Scorpions hunt by night, the prey consis-ting entirely of insects and spiders. They seize the victim with their large claws and tear it to pieces, extracting its body juices. Only if the victim offers resistance will the scorpion sting it. The prey is then slowly eaten, an hour or more sometimes being spent consuming a single beetle. Scorpions can survive long periods of time without eating, and they never drink. They receive their moisture from their food, like most other desert-dwelling animals. Scorpions perform a mating courtship, at which time the female is grasped by the male, who pushes her to a suitable location where he scrapes away the soil with his feet and deposits his spermatophore. Then, still holding the female by her claws, he maneuvers her over the spermatophore so she can sorb it into her body. The young scorpions are born alive, one or two at a time, over a period of several weeks. At birth the baby scorpions ride around on the mother's back. Only after their first molt do they leave the mother and become independent. They may live to an age of 5 years. Look for them on the north bank of the Rogue, where the rainfall is less and the environment is drier.

TICKS

Ticks are one of the most repulsive creatures in the animal world. They are of very little interest to sportsmen, except for the fact that they are bloodsuckers, carriers of disease, and in some cases have caused human deaths. Ticks have a simple life. Females lay 4,000 to 8,000 eggs at random, usually on the ground. The larvae at first hatch have 6 legs ('6-legged larvae') and they climb up plant stems or grasses and wait until an animal passes by, brushing the plant, at which time they transfer to the animal by attaching themselves to the animal's skin, feathers or hair. Then the larva pierces the animal's skin and sucks blood until it is gorged, at which time it removes its head and falls to the ground. The larva then sheds its outer skin and changes into a nymph, which has 4 pairs of walking legs. It then climbs back up the grass and shrubs and duplicates the action of the 6-legged larva. After another molt the 4-legged nymph reaches the 'adult' stage. At this adult stage there is sexual differentiation, the female's body swelling as she feeds and becomes many times larger than the male. Tick bites are troublesome for two reasons: they allow bacteria to enter the skin, causing sores and ulcers; and the tick itself may carry disease-causing organisms which are permitted to enter the host upon the tick's successful completion of its 'bite'. Female ticks die after laying their eggs. Many millions of dollars have been lost in the cattle industry because of disease brought about by ticks. Man himself is also vulnerable, particularly to the disease known as Rocky Mountain Spotted Fever which, fortunately, appears to be absent from the ticks inhabiting the Rogue River Canyon. Another danger to animals and people is the careless removal of ticks, which results in the tick's head being left in the skin. Oil or alcohol applied over the tick will either kill the tick or make it release its hold and drop off. Another method is to apply the lighted end of a cigarette to the back of the tick, which will curl up and drop off. On the river you may use the glowing end of a campfire stick to achieve the same purpose.

SAFETY PREPARATIONS & HOW TO AVOID HAZARDS ALONG THE ROGUE...

Proper footwear is essential when either in the boat or walking along the banks of the river. The correct choice of footwear will help eliminate slipping on rocks and boulders, which can cause serious injuries and spoil your trip. The best footing is obtained by wearing tennis shoes with either crepe or felt soles. You can 'felt' your own shoes by cutting pieces out of indoor/outdoor carpeting and glueing the felt on the bottoms of your tennis shoes with an adhesive like 'Wonder-Glue'. Never let anyone in your party walk barefoot. It is always possible that a piece of glass, buried tin cans or other dangerous objects may be stepped on unknowingly. The sharp, hard rocks can cut like a razor, and the rounded boulders cause stubbed toes and unstable footing. During an emergency situation you cannot take the time to find and put on shoes which may have been removed without anticipating problems.

Adequate cold-weather clothing must be used. You can always take clothing off when you're too hot but you can't put it on if you're too cold if you haven't remembered to bring it with you. It is very helpful to dress in 'Layers' — instead of bringing one heavy jacket, bring one sweater or sweatshirt and a light jacket. On whitewater trips you're very likely to get drenched, so make sure you bring quick-drying clothes. Rayon, nylon or other modern synthetic lightweight clothes are much better than cottons and woolens. Blue jeans are nice and comfortable, but they take hours to dry. Also, remember to always carry lightweight rainclothes with you. Raingear helps stop wind penetration when cold and wet. Bring a long-sleeved shirt or jacket, and long-legged pants to control or prevent sunburn during your trip. A light color or reflective finish is best during sunny daylight trips. Sunglasses and broad-brimmed hat will help control sun glare and excessive heat, thereby preventing headache, sunburned scalps and aching eyes at the end of each day.

Waterproof bags should be included as part of your essential gear. These waterproof containers will keep your sleeping bags, clothing, delicate instruments and other sensitive gear protected and dry. Although garbage bags are waterproof, they will not withstand the physical stress and abuse received during the typical river trip. Good waterproof bags may be purchased in military surplus stores and outdoor supply centers. The bags should be strong enough to serve as cushions, protective padding, emergency fabric and as emergency floating devices.

A waterproof ground cover or tarpaulin is also a very important piece of equipment. It can be used as a ground cloth to sleep on, as a windbreak, as a protective covering at camp to shelter food, firewood and people during rainstorms. It is also useful to keep dew off your sleeping bags in the early morning, as a tablecloth for your meals, as a blanket and a gathering place for groups to discuss the day's experiences, as a water-carrying device and its material can be used to make ropes, emergency repairs, a stretcher, hammock or first-aid bandaging and slings.

A simple camera container can save you hundreds of dollars. An adequate container can range from a simple plastic bag to costly molded camera cases designed to be waterproof and shock resistant. It is important that whatever you choose be waterproof and airtight, so that your camera will be protected and float in case of capsizing. It is always a good idea to tie your camera container to the boat.

The Type 1 PFD (Personal Flotation Device) . . . this 'lifejacket' is designed to keep your head out of water if you are knocked unconscious while in the water. If you are tossed out of your craft, sucked under or forced against rocks, you will feel a lot more secure with these heavy duty lifesaving jackets. They are also designed to prevent your body from slipping out of the jacket, which is a problem with the designs of some of the other lifejacket type flotation vests. Although Type 3 or Type 4 Personal Flotation Devices may look more stylish, they're simply not as safe. If traveling in a large group, bring an extra Type 1 lifejacket along. Kayakers should wear safety headgear, and be sure of their abilities before entering the strong rapids of the Rogue canyon.

A general repair kit should not just include materials for boat repairs, but should also contain miscellaneous other items to handle the repairs of all your other gear. Coleman lanterns should be accompanied by extra mantles, broken fishing rods and reels are an unpleasant (but common) occurence, tent fabric tears are unpleasant. A thousand little problems may occur at some time during your years on the river. Be sure to include at least the following in your kit: some wire, cord, an assortment of nuts and bolts, glue, a needle and thread, pliers, gloves, an axe with a flat end to be used as a hammer, a selection of screwdrivers, a knife, patching material and resin/glue as required. Also include a candle for wax as a lubricant when you need one, vaseline, extra parts for specialized equipment which you take with you on your trip.

Several styles of lifejackets. Type I (recommended) is on right side.

SAMPLE GEAR LIST . . .

FLOATING DEVICE—Some will choose a drift boat, others a raft, kayak or inflatable boat. Whichever floating device you choose, remember to take adequate repairs for your vessel. A completely equipped boat should include the following items:

LIFE JACKET (TYPE1)	BAIL BUCKET (plastic preferred)
AIR-PUMP (for inflatables)	EXTRA OARS or PADDLES
EXTRA OAR-LOCKS	EXTRA TIE-DOWN ROPES
LINING-ROPE (100 ft. length)	SPONGE (to cushion, wipe, bail)

PERSONAL GEAR—each person will have his own preferences, of course, but the following items are recommended to make your trip comfortable and give you maximum versatility:

CHANGE OF PANTS/JEANS/SHORTS	LONG-SLEEVE SHIRT (avoid sunburn)
UNDERWEAR & SOCKS	BATHING-SUIT
SWEATER or SWEATSHIRT	LIGHT JACKET
WALKING SHOES	HAT (broad-brimmed to avoid sunburn)
BOATING SHOES	TOWEL
FLASHLIGHT (with extra batteries)	SOAP
SUNGLASSES	SUN-TAN LOTION (or sunscreen lotion)
INSECT REPELLENT	CHAP-STICK (or other lip protector)
TOILET PAPER (or Kleenex tissue)	TOOTHBRUSH
SHAVING-KIT (or electric-razor)	TOOTHPASTE
POCKET KNIFE	KNIFE/FORK/SPOON/PLATE/CUP
WHISTLE (for safety, better than voice)	(carry these in web-sack for easy cleaning)
CANTEEN (for water)	GLOVES (protect against abrasion/friction)
RAINGEAR (lightweight, waterproof)	DUFFEL-BAG(S) (carry all efficiently)
6' x8' POLYETHYLENE SHEET	

COOKING GEAR—you will probably need to eat during your river trip. Assuming you want to cook your food over an open fire or flame, you will need the following items as well as your food:

FRYING PAN(suitable for size of group)	GRATE (to place over coals)
DUTCH-OVEN (allows amazing versatility)	NESTING POTS/PANS (to suit your needs)
CAN-OPENER (often forgotten)	SPATULA (for mixing/scraping/turning)
LADLE(S) for liquids	LARGE SERVING/COOKING SPOONS
LONG-HANDLED FORKS	ALUMINUM FOIL
PAPER TOWELLING	SALT/PEPPER/SPICES/COOKING OIL
COFFEE POT	PLASTIC BUCKET (mixing batter/salad)
PLASTIC JUICE CONTAINER	SCOURING PAD (a welcome aid)
LARGE METAL CONTAINER TO	SHARP KNIFE (cuts meat/lettuce/etc.)
HEAT DISHWATER (a welcome aid)	DISH TOWELS
TABLE CLOTH	LIQUID DISH SOAP/DETERGENT
GLOVES (to protect hands from heat)	PLIERS (to handle hot metal lids)
MATCHES (or reliable lighter)	FUEL (for stoves if fuel-stoves taken)

CAMPING GEAR—some rugged individualists sleep with no protection or accessories. The rest of us will find the following items to be useful:

TENT	SHOVEL	ICE CHESTS(S)
GROUNDSHEET	LANTERN	AXE (or saw)
FOAM PAD (or air-mattress)	SLEEPING BAG	PILLOW

FIRST-AID KIT—accidents occur when you least expect them. They can be minor cuts, bruises or abrasions—or they can be major, life-threatening crises. BE PREPARED . . . your kit should include the following, but anything additional may also be added:

ADHESIVE TAPE	GAUZE PADS	BAND-AIDS (all sizes)
ASPIRIN (or similar medicine)	TRIANGULAR BANDAGE	SPLINT(S)
SCISSORS	TWEEZERS (or forceps)	RAZOR-BLADE
BLANKET(S)	SALT TABLETS	PETROLEUM JELLY
ABSORBENT COTTON	Q-TIPS (or similar product)	EYE-WASH & CUP
SUNBURN LOTION	ALCOHOL (or antiseptic)	SNAKE-BITE KIT

LODGES AND COMMERCIAL SERVICES ALONG THE ROGUE . . .

BLACK BAR LODGE at mile 8.8

The Main lodge was built in 1935. The work was done by local miners, including Red Keller, who is still employed by the present owners, Bill and Sally Hull. Originally, Mr. & Mrs. Wetherwox operated the lodge, and at that time the clients were packers from the trail who were bringing supplies to the miners. Gradually the miners faded out and the lodge began to service the drift-boat customers brought in by Glen Wooldridge and the Pyle brothers, who were the first river guides. At present the Lodge is solidly reserved by the guide parties and is not available to private groups. The lodge closes about November 10, and re-opens in the spring of the year.

MARIAL LODGE at mile 20.1

This lodge was built by Tom Billings and operated for many years by his daughter Marial, for whom this area was named. Subsequently, the lodge was sold to and improved by Ted Camp, the present owner and operator. Prior to purchasing the Marial Lodge, Ted operated Black Bar Lodge for the Hulls. Like Black Bar, Marial Lodge is solidly booked by guide parties and space is rarely available for others. The lodge closes about November 10 and re-opens in the spring of the year.

PARADISE LODGE at mile 23.6

This lodge was operated for many years by Deek Miller, now retired and living near Agness. The lodge is now owned and operated by Allen and Maryl Boice. Allen was formerly the sheriff of Curry County. In contrast to Black Bar Lodge and Marial Lodge, Paradise Lodge is open all winter. Reservations may be arranged by calling them by radiophone. Dial 247-6249 and ask for Paradise Lodge. A helicopter pick-up is available at either Gold Beach or Medford, or the Allens will arrange for pick-up by jet boat at Gold Beach or Foster Bar. Mrs. Boice is particularly proud of her homemade bread and pies, as well as their fresh vegetables. We can assure you they are excellent. The Mail Boat and Wild Rivers Trips from Gold Beach stop here for lunch for their guests.

HALF MOON BAR LODGE at mile 23.9

This lodge is owned and operated by Bill and Betty Norfleet. The lodge is located off the river and is not visible from the river. There is an airstrip in a large meadow located on the 85- acres of deeded land owned by the Norfleets. Their season runs from May through November, and is open to either down-river traffic (drift-boaters/ trail-hikers, etc.) or they will pick you up at Foster Bar and jet-boat you back to the lodge. It is also possible for guests to be flown in from either Grants Pass or Gold Beach. This lodge is located at one of the finest fishing areas of the river. Half Moon Bar gets its name from the crescent shape of the bar itself.

CLAY HILL LODGE at mile 28.5

This lodge is owned by Tom Staley. It caters primarily to guided parties, but occasionally reservations may be obtained by calling Paradise Lodge by radiophone, or through Court Boice (see Peyton Place Lodge, below).

PEYTON PLACE LODGE at mile 31.0

Also called Wild River Lodge, this lodge is owned by Tom Staley, operated by Court Boice. It also caters primairly to guided parties, but occasionally reservations may be obtained by calling Court Boice by radiophone. Dial 247-6022 and ask for Peyton Place Lodge.

SELECTED READINGS

Atwood, Kay, *Illahe,* The story of settlement in the Rogue River Canyon, Medford, Oregon, Gandee Printing, 1978.

Beckham, S. D., *Tall Tales from Rogue River — The Yarns of Hathaway Jones,* Bloomington, Indiana, Indiana University Press, 1974.

Collins, Robert and Nash, Roderick, *The Big Drops, Ten Legendary Rapids,* San Francisco, California, Sierra Club Books, 1978.

Kuhne, Cecil, *River Rafting,* Mountain View, California, World Publications, 1979.

McGinnis, William, *Whitewater Rafting,* New York, Quadrangle-New York Times, 1975.

Pringle, Laurence, *Wild River,* Philadelphia, Lippincott, 1972.

Schwind, Dick, *West Coast River Touring: Rogue River Canyon and South,* Beaverton, Oregon, Touchstone Press, 1974.

Tejada-Flores, Lito, *Wildwater: The Sierra Club Guide to Kayaking Whitewater Boating,* San Francisco: Sierra Club Books, 1978.

Yocom, Charles and Brown, Vincent, *Wildlife and Plants of the Cascades,* Healdsburg, California, Naturegraph Publishers, 1971.

Quinn, James W., Quinn, James M., and King, James G., *Handbook to the Deschutes River Canyon,* Medford, Oregon, Commercial Printing Company, 1980.

Quinn, James W., Quinn, James M., and King, James G., *Handbook to the Illinois River Canyon,* Portland, Oregon, Printers Northwest, 1979.